Mathematics for College Physics

Biman Das

The State University of New York at Potsdam

PEARSON

Prentice Hall

Upper Saddle River, NJ 07458

Library of Congress Cataloging-in-Publication Data
Das, Biman
  Mathematics for college physics / Biman Das.
     p. cm.
Includes bibliographical references.
  ISBN 0-13-141427-5 (pbk.)
  1. Mathematics. 2. Mathematical physics. I. Title.
QA39.3.D37 2004
510--dc21

                                        2003014878

Associate Editor: *Christian Botting*
Senior Editor: *Erik Fahlgren*
Editor in Chief, Science: *John Challice*
Production Editor: *Donna Young*
Executive Managing Editor: *Kathleen Schiaparelli*
Manufacturing Buyer: *Alan Fischer*
Manufacturing Manager: *Trudy Pisciotti*
Copy Editor: *Write With, Inc.*
Art Director: *Jayne Conte*
Cover Designer: *Kiwi Design*
Editorial Assistants: *Nancy Bauer, Andrew Sobel*

© 2004 Pearson Education, Inc.
Pearson Prentice Hall
Pearson Education, Inc.
Upper Saddle River, NJ 07458

Printed in the United States of America
10 9 8 7 6 5 4 3 2 1

ISBN 0-13-141427-5

Pearson Education LTD., *London*
Pearson Education Australia PTY, Limited, *Sydney*
Pearson Education Singapore, Pte. Ltd
Pearson Education North Asia Ltd, *Hong Kong*
Pearson Education Canada, Ltd., *Toronto*
Pearson Educacion de Mexico, S.A. de C.V.
Pearson Education—Japan, *Tokyo*
Pearson Education Malaysia, Pte. Ltd

Dedicated to my mother Mrs. Renuka Rani Das

Table of Contents

Chapter 5
Trigonometry: A Powerful Tool for Solving Real-World Problems 116

Right-Angle Trigonometry 117

Non-Right-Angle Trigonometry 140

Chapter 6
Vectors: Tracking the Direction of Quantities 142
Introduction to Vectors 142

Vector Addition and Subtraction 151

Chapter 7
How to Be Successful in a Physics Laboratory: Analysis of Data, Curve Fitting, Probability, and Statistics 166

Preface

This book is written for students who plan to take or who are presently taking an algebra-trigonometry-based physics course. The book will develop mathematical skill, provide the students with the competence to use mathematics, and serve as a mathematical resource. Students will find how mathematics is directly applied to physics. Students who have not recently taken a course that required mathematical skill often find themselves unable to keep pace in an introductory physics course because they are not familiar with the necessary mathematical tools. This book is written with these students in mind. Students taking other physics and science courses will also find the book useful.

The book consists of seven chapters. Chapter 1 introduces students to physics and mathematics and discusses the role of mathematics in physics. It also discusses how to deal with the math anxiety that many students have, as well as how to develop good study habits. Chapter 2 describes scientific notation, units, and dimension of physical quantities. Chapters 3 through 7 describe the mathematical concepts and tools including algebra, geometry, trigonometry, vector, and statistics, with selected examples and real-world problems. Chapter 7 covers statistics and elaborates on the techniques of analyzing experimental data; it provides tips on how to be successful in a physics laboratory.

The book emphasizes primarily the use of mathematical techniques and mathematical concepts in physics and does not go into their rigorous developments. Wherever possible, I have included examples and problems from physics so that students see how the concept of mathematics is directly applied to physics. The book is written so that students can study it to develop mathematical skills and can practice solving problems. There are worked-out examples and exercises in almost every section. The book can also be used as a text for a one- to two-credit course (dependent upon the depth of the coverage), a remedial mathematics course, or a physics recitation course for students taking an introductory physics course. Both the publisher and I realize the necessity of the book for introductory physics students.

I would like to thank Dr. David Reid (East Michigan University), Dr. Ronald J. Bieniek (University of Missouri–Rolla), Dr. Anthony Pitucco (Pima Community College), Dr. Jenny Quan (Pasadena City College) and Dr. Dennis Rioux (University of Wisconsin–Oshkosh) for their reviews and suggestions. Special thanks are due to Dr. Michael B. Ottinger (Missouri Western State College) for his detailed comments and suggestions for improvements. I thank Mr. Christian Botting, Associate Editor, Physics and Astronomy, Prentice Hall, for his constant support and cooperation throughout the entire process. I also thank Donna Young, Production Editor, Prentice Hall, for her cooperation. I thank my wife Indrani Das and my daughters Debapria Das and Deea Das for their cooperation and support without which I could not get any free time to write the book. I would appreciate any constructive comments about the book.

Biman Das

To the Students

The primary purpose of this book is to develop mathematical skills that will allow you to concentrate on the physics in a physics course. You should find the book useful in other science courses that involve problem solving, and you can use the book as a mathematics resource book. The book is not meant to be a substitute for standard mathematics courses that may be prerequisites for an introductory physics course. The mathematics discussed is comparable to what you will find in a typical mathematics text. The emphasis is on enhancing those mathematical skills frequently used in a physics course. One objective of the book is to show you how the concepts of mathematics are directly applied and related to physics and real-world problems.

For many of you, college level physics is different from most other courses. The emphasis is on understanding the physical concepts, deriving the physical laws or principles using mathematics, and applying this knowledge in solving a wide range of problems. There may be several obstacles to understanding physics and mastering problem solving. However, the primary factor that leads to success in physics and problem solving is the competence to use mathematics with ease and accuracy. My hope is that the book will make you stronger in mathematics so that your learning experience in physics will be beautiful and rewarding. The beauty of physics cannot be enjoyed if you are distracted and inhibited by a less-than-adequate mastery of basic mathematical skills.

For your convenience, I have included important conversion factors, a list of fundamental constants, useful data, and other information in the appendices. I have used boxes throughout the book to draw your attention to important concepts or formulas. I have also used the boxes to inform you of students' common mistakes.

Read Chapter 1 to learn about fun with physics and mathematics. Learn also how to develop good study habits and how to deal with math anxiety. Then begin Chapter 2. Good luck!

Biman Das

Chapter 1
Fun with Physics and Mathematics

The fundamental principle of natural philosophy (physics) is to attempt to reduce the apparent complex physical phenomena to some simple fundamental ideas and relations. — Einstein and Infeld

1.1   How Physics Fits into Science

The word science was taken from the Latin word "scientia," meaning "to know." Scientists currently define science as the systematic attainment of *knowledge* derived from observation, study, and experimentation in order to understand nature and its principles. Aristotle (384–322 B.C.) was one of the first persons to study science in a systematic manner. The greatest advancement of science came about at the beginning of the sixteenth century when people, such as Galileo (1564–1642), started looking for the relationships between physical variables and tested their ideas through experimentation.

As science grew, it was divided into two major groups: the biological sciences and the physical sciences. Biological sciences represent an organized study of living matter while physical sciences represent an organized study of inanimate matter and energy. The physical sciences include physics, geology, chemistry, astronomy, and meteorology [Figure 1]. Geology primarily deals with the science of earth. Chemistry deals with the structure, properties, and conversions of matter. Astronomy deals with the science of celestial bodies including planets, stars, and galaxies. Meteorology is the science of weather. It is believed that nature follows certain principles and there exists an order among a wide variety of natural phenomena. The primary role of physics is to *discover the fundamental principles of nature*. Physics holds the central position in the physical sciences, since it is concerned with the fundamental aspects of matter and energy as well as how they interact to make the universe function. The fundamentals of physics, on the one hand, include concepts and mathematics, and on the other hand, the practical applications of technology.

The field of physics is divided into the areas of mechanics, waves, optics, thermodynamics, and electricity, which make up the *classical* physics developed between the sixteenth and nineteenth centuries by Galileo, Newton, Faraday, Maxwell, and others. Physics also includes modern topics developed in the twentieth century, such as relativity, quantum physics, and quantum field theory, as well as their appropriate applications in condensed matter, atomic and nuclear phenomena, elementary particles, and astrophysics.

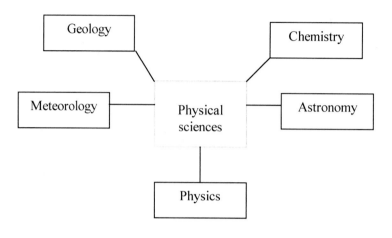

Figure 1

The Newtonian mechanics that holds perfectly well for ordinary moving objects was found to be inadequate for objects moving close to the speed of light and has been replaced by the special theory of relativity of Einstein (1905). For extremely small objects (comparable to the size of an atom or less), Newtonian mechanics has been superseded by quantum mechanics, as developed in the early twentieth century by Schrödinger, Heisenberg, and others. In the middle of the twentieth century, quantum field theory was developed that combined relativity and quantum principles. The great four areas of physics are shown in Figure 2.

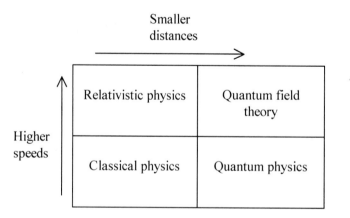

Figure 2

A solid foundation in physics is needed to fully understand any natural science. All the complexity and variety that you see in the world around you is a manifestation of a few fundamental principles of physics. It should be fun for you to know why the universe behaves the way it does. For many people, physics is a mechanical process of collecting facts. I do not think so. Physics, like art, poetry, and painting, is a creative activity that engages the emotion and the intellect. Physics is not a copy of nature, but a re-creation of nature by the act of discovery.

Physicists love physics because physics is beautiful. Physics shows us the elusiveness of the ultimate truth. Once you really understand physics, it will show you the same aesthetic qualities that you find in an exquisite work of art, a beautiful poem, or a pleasant piece of music.

The Book of nature is written in mathematical characters. — Galileo Galilei

1.2   Mathematics: The Language of Physics

Physics is a fascinating science, and make no mistake about it—physics is not mathematics. *Mathematics is the study of numbers and the geometry of objects.* It is based on logic and the rules of counting. *Physics is the study of the rules that govern nature and of the fundamental processes occurring in nature.* Observation, careful experiments, and measurements comprise one aspect of physics. The other aspect of physics is the creation or invention of theories that explain the observation. Theoretical physics depends heavily on mathematics. You should find it very interesting that the principles and laws of physics can be expressed in terms of beautiful mathematical equations. Trying to study physics while totally avoiding mathematics is practically impossible. It would be the equivalent of studying Shakespeare while avoiding the written word. It simply cannot be done meaningfully.

People have noticed such a connection between physics and mathematics from ancient times. In ancient Greece, Pythagoras first noticed that pleasing notes were produced when the length of a string was changed by whole numbers relative to its original length. In the seventeenth century, Galileo discovered that the distance a freely falling object fell could be expressed by an equation that includes the square of time. The relationship between the cause (force) and effect (acceleration) for a motion as expressed by Newton's second law of motion is a simple mathematical statement. Newton developed calculus (which was also developed by Leibniz), a powerful mathematical tool, to probe the changes in nature. For example, the position, velocity and acceleration of a mass attached to a spring at any time can be determined using calculus. In fact, all basic laws of physics can be expressed in beautiful and essentially simple mathematical forms. But why nature follows such beautiful mathematical equations is a mystery. I do not think anybody knows the answer to this question.

But we are fortunate that nature has allowed us to explore its hidden truth and beauty through the use of mathematics developed by humans.

> *The most fundamental ideas of science are essentially simple, and, as a rule, can be expressed in a language comprehensible to every one.* — Albert Einstein

1.3   Why Does Physics Appear to Be Difficult?

It is not uncommon to hear from students that physics is difficult. There are some genuine reasons for this feeling:

- First of all, most physics courses rely heavily on mathematics for the understanding of physical processes. Real-world problem solving in physics also requires knowledge of mathematics. If you are anxious about your skills in mathematics, you will find physics to be difficult.

- Although the most fundamental ideas or concepts of physics are essentially simple, sometimes physics instruction is not always engaging.

- In introductory physics courses, you need to learn a great deal of course material within just two semesters. You must spend a lot of energy and effort to solve homework problems, read the text, attend laboratories, and take quizzes and exams. It is to be expected that you might need additional help from the instructor outside of the classroom. You may fail to get the additional help due to time conflicts in your busy schedule or with the available hours of the instructor or teaching assistant.

- Finally, getting a good grade in a physics course is a challenge, even if you do attend classes and work regularly.

The point is how well you are prepared to meet the challenge. The obstacles may be several, but the primary factor that leads to success is your competence in using mathematics in physics with ease and accuracy. Meeting the math prerequisite for a physics course may not be enough. Students who have not recently taken a course that required mathematical skill often find themselves unable to keep pace in a physics course because of a lack of familiarity with the necessary mathematical tools.

Do not worry about your difficulties in mathematics, I can assure you that mine are still greater. —Albert Einstein

1.4   Math Anxiety and How to Deal with It

Feeling nervous before an exam is not uncommon among students. Feeling slightly nervous when it comes to math is also not uncommon. However, if you really panic because you need to do math, it is likely that you have math anxiety. You need to ask yourself what are the causes of this anxiety and find out the root of these causes. Admit that math is sometimes not obvious and you, like others, need to work hard to understand. I have the following suggestions to get rid of this anxiety:

- *Practice math to develop your math skills.* This is the first and most important advice for dealing with this anxiety. Once you have adequate math skills, you can concentrate more on the physics in a physics course. Talk with your physics instructor to find out what would help you to develop appropriate skills in mathematics.

- *Join a physics team involving three to four students* from the same physics class and *meet regularly* to discuss math as well as physics problems and concepts. You will find that you can develop your math skills and learn physics much easier in a group than on your own. Research has shown that students can learn well from their peers.

- *Choose your teammates carefully* to appropriately mix good work and fun. *Choose a quiet and comfortable regular meeting place.*

- *Participate actively in the group* and *in the class.* The more you discuss a problem with others, the easier it will be to recall. If you are lost, ask a team member to clearly explain. Do not copy from others.

- *Keep your body healthy.* Healthy bodies can think quicker and can work harder.

- *Take a break and relax if you are frustrated or nervous.* Learn and practice the techniques recommended for relaxing and beating stress. You will be able to concentrate more in math, if you are less stressed.

- *Do not give up, get help.* There is no shame in getting help. If you need a remedial course in mathematics, take it. Many institutions offer a *physics recitation* course parallel to an introductory physics course. The physics recitation course is a good chance to practice problems and develop appropriate math skills.

Remember, math anxiety can certainly be remedied if you make a serious effort. You can master mathematics if you work at it with caution and care.

When I really understand something, it is as if I had discovered it myself. —Richard Feynman

1.5   Developing a Good Study Habit: Studying for Success

At the college level it is essential to study properly for success. First of all, the course material may be difficult to understand in one sitting. Second, you may not have enough time available between classes and extracurricular activities. Therefore, developing a good study habit and proper study technique are necessary steps for your success. Use the following tips to help you to develop a proper study technique and good study habit.

- *Plan your study wisely.* That is, make a realistic timetable that will include your study times as well as times for other activities that you need and want to do. The plan must be *realistic* so that you can stick to it throughout the course. *Maintain a calendar* to record important commitments such as tests, exams, due dates of papers, etc.

- *Be a good reader.* Reading text is one critical part of the learning process, because all material that is included cannot be presented in the lecture. Use a highlighter to highlight the *main ideas*. Use a pencil to *take notes* in the margins. Always *think* about what you have read. Work the example problems of the section that you have read. Worked-out examples in the texts are usually good, and these may represent the standard of your test or exam.

- *Come to the class prepared.* That is, study the previous lecture and at least the outlines of the next lecture from the text before you attend the next class. Concentrate on the lecture. Concentration is a skill that can be improved by practice

- *Interact with your instructor* in the class. It is a good habit to sit in the class near the center of the front row so that instructor can clearly see your activities and you will have the minimum chance of being distracted.

- *Take good notes in class.* Your notes must include *main ideas* along with *supporting details* and *examples*. Rewrite your notes and create flash cards with important facts, laws, and vocabulary. Keep the flash cards handy and look at them as often as needed.

- *Join a study group* that can appropriately mix good work and fun. Choose your partners in the group carefully. Have discussion with others in the group and practice problems by collaborative efforts. Knowing or understanding a topic or a problem may not be enough, only *practice* can make you perfect in a test or exam.

- *Prepare well for the tests or exams, starting at least a week before.* Preparation is a *continuous daily activity* and not a task to be started the night before a test or an exam. As you study for the test or exam, ask yourself what questions or types of problems could be asked.

- Take a break from your study when needed. Also, take a break if it helps you increase your concentration on your studying. Do not feel guilty if you need a break.

- *Get help when you need it*. Meet your instructor outside of class to ask questions about concepts, walk through homework assignments, and to review for tests and exams. Do not feel guilty if you need to see your instructor quite often during office hours. Discuss old tests, quizzes, exams, or practice problems with your instructor. Your library is a useful resource for books and other academic material. Take advantage of this. Practice old tests, quizzes, or exams, if available. Important study guide material or other resources for the course are often on reserve at the library. Use them as often as necessary.

- *Keep your body healthy and learn how to relax*. Stress can have negative effects on you and your thoughts. Too much stress can cause anxiety or even depression.

You do not know anything until you have practiced. —Richard Feynman

1.6   Solving Problems in Physics

To learn physics, you must practice by solving problems. Your ability to solve problems is one of the measures of whether you truly understand physics. To solve problems in physics, you must have the necessary mathematical skills. Remember that many problems can be solved by more than one method. Research has shown that students can solve problems well by collaborative efforts. Although there is no fixed recipe for solving problems in physics, you should find the following steps useful:

- *Read the problem carefully*. It is important to understand the meaning of each word in the problem. Recognize the meaning of different terms (e.g., *at rest* means that the initial velocity $v_0 = 0$; *a smooth, flat surface* means a horizontal and frictionless surface; moving at a *constant velocity* means that the acceleration $a = 0$). Read the entire problem. Texts often provide important hints at the end of a multiple part problem. If you only read part (a), you will miss the hints.

- *List data*. Some of the listed data may be redundant in some problems. For example, the mass of an object will not be necessary to determine its acceleration from the knowledge of time and its initial and final velocities.

- *Check units in the data and convert units where necessary*. Be careful about the units. Convert to SI units whenever possible. Experience shows that students frequently get wrong results for problems either because the units are wrong or they made mistakes during unit conversions.

- *Draw a sketch or diagram of the problem wherever appropriate*. A simple sketch that depicts the spatial arrangements described in a problem greatly aids understanding of the problem. Some problems may require the adoption of a *sign convention* (e.g., upward direction as positive and downward direction as negative). *Stick with the same sign convention* for the whole problem. Some problems may require *setting up a convenient coordinate system* for the solution. In most situations, the Cartesian coordinate system (with horizontal direction as the *x*-axis and vertical direction as the *y*-axis) is suitable. However, for the motion of an object on an inclined plane the most convenient axes are usually the x-axis along the plane of inclined surface, and the y-axis perpendicular to the plane of the inclined surface.

- *Devise a strategy or plan for the solution.* Determine which physical principles and equations of physics can be applied. That is, think about the physics behind the problem and then relate the situation to physical principles and equations that describe the physics (e.g., if the problem involves force and acceleration, then it involves Newton's second law; if the problem involves a collision between two objects, it relates to the conservation principle of linear momentum).

- *Translate principles into the language of equations.* For example, if you are dealing with a kinematics problem, and the initial and final velocities and time are given, use the equation $v = v_0 + at$.

- Discuss with others if the problem needs to be solved by collaborative efforts.

- *Substitute given quantities into equations,* and use your knowledge of mathematics to solve them or to do calculations to get the result. *Be aware of the common mathematical mistakes* (e.g., $1/a + 1/b$ is not equal to $1/(a + b)$; $x^2 + y^2$ is not equal to $(x + y)^2$).

- *Report the result in scientific notation keeping the proper number of significant figures with the proper unit.*

- *Evaluate your answer and determine whether the result makes sense* (e.g., find whether the result has the expected magnitude, dimension, unit, sign, etc.). In some situations, only one of the two mathematical solutions is physically acceptable. If you solve equations to get the result, substitute the results in the equations to verify whether they satisfy the equations. If the problem can be solved by an alternative method, you may also quickly check the result using the alternative method.

Example 1.6.1

Determine the time taken by a ball to roll with a constant velocity of 0.24 meters per second (m/s) across a 4.0-meter (m) long table.

The problem may seem simple, but the main purpose of this problem is to take you through the steps of good problem solving.

1. Read the problem.   Note that the phrase *constant velocity* means there will be no acceleration ($a = 0$).

2. List data.

$v = 0.24$ m/s
$x_i = 0.0$ m
$x_f = 4.0$ m
$a = 0$
$\Delta t = ?$ ← what we want to know

3. Check units.   In this case each unit is in standard form so no conversions are necessary.

4. Draw a sketch.

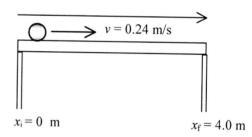

$x_i = 0$ m $x_f = 4.0$ m

5. Devise a strategy or plan for the solution.   This is a kinematics problem that can be solved using the definition of constant velocity.

6. Translate principles into the language of equations.

$$v = \frac{x_f - x_i}{\Delta t}$$

7. Substitute given quantities into equations and solve for unknown.

(i). Insert numeric values:

(ii). $0.24 \text{ m/s} = \dfrac{4.0 \text{ m} - 0.0 \text{ m}}{\Delta t}$

(iii).   Divide   by   0.24   m/s:

$\Delta t = \dfrac{4.0 \text{ m}}{0.24 \text{ m/s}} = 16.667 \text{ s}$

8. Report the result in scientific notation with significant digits.

Use two significant digits, since all initial data are given to two significant digits:
$\Delta t = 1.7 \times 10^{1} \text{ s}$.

9. Evaluate your answer.

Units: Seconds — correct dimensions, since we were looking for a time.
Magnitude: Since traveling about 1/4 m/s and needed to travel 4 m, 17 s seems reasonable.
Sign: Positive, as expected.
Does the solution answer the question posed in problem? Yes.

Example 1.6.2

Draw a simple sketch to illustrate each of the following situations that you may find in a physics problem:

(a) A box resting on an inclined surface.

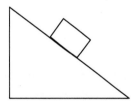

(b) A ball rolling down an incline.

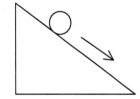

c) An object is pulled at an angle of 45° above the horizontal on a horizontal surface.

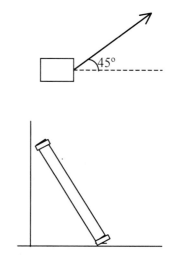

(d) A ladder is leaning against a wall.

(e) An airplane flies 50 mi east and then makes a left turn at an angle of 30° north of east and flies another 100 mi.

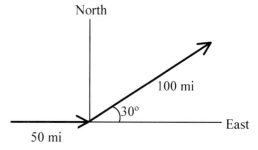

(f) Collision of two solid balls moving in the same direction before, during, and after collision (moving in the opposite direction).

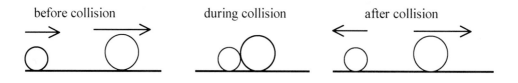

Chapter 2
Scientific Notation, Units, and Dimensions of Physical Quantities

2.1   Powers of 10 and Scientific Notation

In physics, you will deal with numbers that are extremely small and extremely large. For example, the lifetime of an unstable particle, called the muon, is 0.0000022 second, which is very small, whereas the mass of earth is 597 000 000 000 000 000 000 000 0 kilograms, which is very large. Such extremely small and large numbers take space and time to write. Scientists express such very small or large quantities in *scientific notation* using only the significant figures. Note the following notation:

$10^0 = 1$ $10^0 = 1$

$10^1 = 10$ $10^{-1} = \dfrac{1}{10} = 0.1$

$10^2 = 10 \times 10 = 100$ $10^{-2} = \dfrac{1}{10 \times 10} = 0.01$

$10^3 = 10 \times 10 \times 10 = 1000$ $10^{-3} = \dfrac{1}{10 \times 10 \times 10} = 0.001$

and so on.

In *scientific notation*, a number is written as the product of a number from 1 through 10 and an integer power of 10. That is, written in the scientific notation, every number has the following form:

(A decimal between 1 and 10) $\times\ 10^{\text{An integer exponent}}$

Thus, expressed in scientific notation,
the lifetime of muon $= 2.2 \times 10^{-6}$ s
the mass of earth $= 5.97 \times 10^{24}$ kg.

Exercise
1. Express the following numbers in scientific notation:
    a) 272000   b) 0.000000534   c) 324×10^2 d) 128×10^{-6}
    e) 0.0000054

2.2   Significant Figures and Rounding Off

The numbers dealt with in mathematics are exact. For example, when a mathematician writes 1, it means 1.000.... That is, 1 means exactly 1. The situation is different in physics. Many of the numbers dealt with in physics come from the measurements of physical quantities. The measured numbers of physical quanatities are usually not exact. Exact numbers do not have any uncertainty or error. Measured numbers usually have some degree of uncertainty or error. For example, suppose the measured distance by a centimeter ruler is 2.53 cm. Here, 3 is an estimate meaning that the distance is not less than 2.5 cm or more than 2.6 cm. The 2, the 5, and the 3 in the measurement are called *significant figures*, because they each provide reliable information about the length of the object. A centimeter ruler measurement is good to three significant figures, whereas a precise micrometer can yield a more precise value of the measured length, although that will still not be exact. The number of *significant figures* in a physical quantity is equal to the number of digits that are estimated or known with some reliability. Also, in numerical calculations using a calculator, the calculator may give a result that has many figures, which are not meaningful.

Use the following rules to determine the significant figures:

- Zeros at the beginning of a number are not significant. For example, in 00056, the zeros are not significant.

- Zeros within a number are significant. For example, in 5.04, the zero is significant.

- Zeros at the end of a number after the decimal point are significant. For example, in 54.00, the zeros are significant.

In whole numbers without a decimal point that end in one or more zeros (for example 500 kg), the zeros may or may not be significant. Scientific notation allows the number of significant figures to be clearly expressed. For example, consider the number 23,400. If the number is known up to three significant figures, we write 2.34 X 10^4, whereas if the number is known up to four significant figures, we write 2.340 X 10^4.

Use the following rules to round off numbers:

1. If the first digit to be dropped is less than 5, leave the preceding digit as is (e.g., 2.162 becomes 2.16 rounded to three significant figures).

2. If the first digit to be dropped is 5 or greater, increase the preceding digit by one (e.g., 2.167 becomes 2.17 rounded to three significant figures).

Example 2.2.1

Express the following numbers in scientific notation with three significant figures:

  a) 0.0241

  b) 79,241

  c) 0.000452

The numbers in scientific notation with three significant figures are as follows:

  a) $0.0241 = 2.41 \times 10^{-2}$

  b) $79,241 = 7.92 \times 10^{4}$ (Note: 7.9241 has been rounded off to 7.92.)

  c) $0.000452 = 4.52 \times 10^{-4}$

Example 2.2.2

Express the following numbers as their decimal equivalents.

  a) 49.6×10^{-2}

  b) -6.50×10^{-4}

  c) 0.4291×10^{3}

  d) 1.21×10^{0}

The decimal equivalents are as follows:

  a) $49.6 \times 10^{-2} = 0.496$ (by moving the decimal point two places left)

  b) $-6.50 \times 10^{-4} = -0.000650$ (by moving the decimal point four places left)

  c) $0.4291 \times 10^{3} = 429.1$ (by moving the decimal point three places right)

d) $1.21 \times 10^0 = 1.21$

Example 2.2.3

The nearest star to the earth is Alpha Centauri whose distance is approximately 25 700 000 000 000 miles. Express the result in scientific notation with three significant figures and four significant figures.

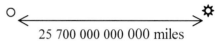

25 700 000 000 000 miles

$25\ 700\ 000\ 000\ 000$ miles $= 2.57 \times 10^{13}$ miles, written in scientific notation with three significant figures. Expressing with four significant figures, we find that the result is 2.570×10^{13} miles. Note that, in the latter case, the zero after the decimal point at the end of 2.570 is significant.

Example 2.2.4

In each of the following expressions, rewrite each factor and term in scientific notation and express the final result of the computation in scientific notation with three significant figures:

a) 51.5×0.0000521

First, express each number in scientific notation.

51.5×0.0000521

$= (5.15 \times 10^1) \times (5.21 \times 10^{-5}) = (5.15 \times 5.21) \times 10^{-4}$

$= 26.8315 \times 10^{-4}$

The result is not correctly written in scientific notation. To get the decimal place in the right place, we move one decimal place to the left, increasing the exponent by + 1. Thus, the answer in scientific notation is 2.68315×10^{-3}. Keeping the result up to the three significant figures, we round off the result and rewrite it as 2.68×10^{-3}.

b) $\dfrac{2348}{83.1 - 91.8}$

First, express each number in the fraction in scientific notation.

$$\frac{2348}{83.1-91.8}=\frac{2.348\times10^3}{8.31\times10^1-9.18\times10^1}=-2.70\times10^2$$

Here the result is shown in scientific notation up to the three significant figures.

The result of a measured quantity, or of a problem, should be reported with the proper number of significant figures. Use the rules (mentioned in the accompanying boxes) to determine the number of significant figures to be kept after adding, subtracting, multiplying, and dividing numbers.

When adding or subtracting numbers, leave the same number of decimal places in the answer as there are in the number with the least number of significant figures.

Add the lengths of two tables of lengths 6 ft and 4 ft, 3 in. Since the first table does not specify the inches, we must round our final answer to 10 ft. This is because, if no inches are specified, "6 ft" means "anywhere between 5 ft, 6 in to 6 ft, 6 in."

For example, suppose you need to add 702.3, 124.0, 42.06 and 0.0542. The result from your calculator is 868.4142. Now, look at your data and observe that in two of the numbers being added it is not clear what the figure in the second decimal place is. The result must be rounded off to one decimal place by dropping 142. Hence, the answer with proper numbers of significant figures is 868.4.

When multiplying and dividing numbers, leave as many significant figures in the answer as there are in the number with the fewest number of significant figures.

Usually, multiplication and division produce many more figures that are significant; thus, the result must be rounded off. For example, suppose the measured length and width of a plate are 2.64 cm and 1.18 cm, respectively. Therefore, the area = 2.64 cm X 1.18 cm = 3.1152 cm^2. The result appears to have five significant figures, although the original quantities have three each. Therefore, the result can have only three significant figures and is 3.12 cm^2 (rounded to the second decimal place).

The next example also illustrates how to determine the correct number of significant figures in addition, subtraction, multiplication, and division.

Example 2.2.5

Determine the correct number of significant figures in the following operations:

a) $(5.32 \times 10^{-4}) + (7.4 \times 10^{-4})$

b) $(3.42 \times 10^{-6}) - (2.8 \times 10^{-6})$

c) $(5.212 \times 10^{3}) \times (7.8 \times 10^{5})$

d) $(8.2 \times 10^{5}) \div (7.62 \times 10^{2})$

a) $(5.32 \times 10^{-4}) + (7.4 \times 10^{-4}) = (5.32 + 7.4) \times 10^{-4} = 12.72 \times 10^{-4}$

The final answer can have only the same number of decimal places in the answer as there are in the number with the least number of decimal places. The number with least number of decimal places in this problem is 7.4×10^{-4}. Since 7.4×10^{-4} is expressed to the one decimal place, the result is 12.7×10^{-4}. To express the result in scientific notation, move the decimal point to the left one position, increasing the exponent by +1. Thus, the result is 1.27×10^{-3}.

b) $(3.42 \times 10^{-6}) - (2.8 \times 10^{-6}) = (3.42 - 2.8) \times 10^{-6} = 1.62 \times 10^{-6}$

The number with the least number of decimal places is 2.8×10^{-6}, which has two significant figures. The final answer can have only two significant figures and, hence, the result is 1.6×10^{-6}.

c) $(5.212 \times 10^{3}) \times (7.8 \times 10^{5}) = (5212 \times 780000) = 4065360000 = 4.1 \times 10^{9}$, rounded up to two significant figures.

d) $\dfrac{8.2 \times 10^{5}}{7.62 \times 10^{2}} = \dfrac{820000}{762} = 1076.1155 = 1.1 \times 10^{3}$, rounded up to two significant figures.

Exercises

2. a) Express 200 kg in scientific notation with two significant figures.

b) Express 200 kg in scientific notation with three significant figures.

3. Round off the results in scientific notation to the proper number of significant figures:

a) Multiplication: 2.4 m X 3.65 m

b) Division: $\dfrac{7.24\,m}{1.116\,s}$

c) Addition: $8.1 \times 10^4\,m + 9.1 \times 10^3\,m + 0.008 \times 10^6\,m$

2.3   Physical Quantity, Units, and Prefixes

In physics, you will deal with different physical quantities, such as length, mass, force, and others. A number and a unit express a *physical quantity*. There are two types of physical quantities: *base quantities* (or fundamental quantities) and *derived quantities*. Derived quantities can be expressed in terms of base quantities. In your study of physics, you will come across many different units. Some are familiar to you (such as inches, gram, etc.) and some may be unfamiliar (such as newtons, pascals, etc.). Use the *metric system* of units whenever possible in solving problems. In the metric system, you multiply (or divide) by a power of 10 when you change from a larger (or smaller) unit to a smaller (or larger) unit. The following diagram is illustrative:

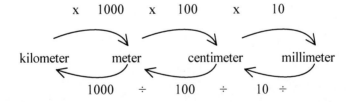

The SI unit (International System of Units) was adopted in 1960 as standard units to express physical quantities. The SI units of length, mass (i.e., the amount of matter), and time are *Meter* (m), *Kilogram* (kg), and *Second* (s), respectively. These units are also called MKS units. The SI units of base and derived quantities are shown in Table 2.1 and Table 2.2, respectively.

Table 2.1: International System (SI) of Units for Base Quantities

Quantity	Unit	Abbreviation
Length	meter	m
Mass	kilogram	kg
Time	second	s
Electric current	ampere	A
Temperature	kelvin	K
Luminous Intensity	candela	cd
Amount of substance	mole	mol

Table 2.2: International System (SI) of Units for Derived Quantities

Quantity	Unit	Abbreviation
Angle	radian	rad
Solid angle	steradian	sr
Area	square meter	m^2
Volume	cubic meter	m^3
Density	kilogram per cubic meter	kg/m^3
Velocity	meter per second	m/s
Acceleration	meter per second squared	m/s^2
Angular velocity	radian per second	rad s
Angular acceleration	radian per second squared	rad/s^2
Force	newton	$N\ (kg{\cdot}m/s^2)$
Torque	newton-meter	$N{\cdot}m$
Work, energy, heat	joule	$J\ (N{\cdot}m)$
Power, radiant flux	watt	W
Pressure, stress	pascal	$Pa\ (N/m^2)$
Frequency	hertz	Hz (1/s)
Electric charge	coulomb	$C\ (A{\cdot}s)$
Voltage, electromotive force	volt	V (W/A)
Electric field	volt per meter	V/m
Electric resistance	ohm	Ω (V/A)
Capacitance	farad	$F\ (A{\cdot}s/V)$
Magnetic flux	weber	$Wb\ (V{\cdot}s)$
Inductance	henry	$H\ (V{\cdot}s/A)$
Magnetic flux density	tesla	$T\ (Wb/m^2)$
Luminous flux	lumen	Lm
Luminance	lux	lx
Activity (radioactive)	becquerel	$Bq\ (s^{-1})$
Absorbed dose (of radiation)	gray	Gy (J/kg)
Dose equivalent (dose equivalent index)	sievert	Sv (J/kg)

Prefixes

In the metric system, the larger and smaller units are defined in certain powers of 10 from the standard unit to make the conversion easy. The *prefixes* can be used to express units of any physical quantity in the metric system. The prefixes, their symbols, and their meanings are listed in Table 2.3. Table 2.4 lists the most common prefixes used in introductory physics.

Table 2.3: Prefixes and Their Meanings

Prefix	Symbol	Meaning	
peta	P	10^{15}	=1000 000 000 000 000
tera	T	10^{12} =	1 000 000 000 000
giga	G	10^{9} =	1 000 000 000
mega	M	10^{6} =	1 000 000
kilo	k	10^{3} =	1 000
deci	d	10^{-1} =	0.1
centi	c	10^{-2} =	0.01
milli	m	10^{-3} =	0.001
micro	μ	10^{-6} =	0.000 001
nano	n	10^{-9} =	0.000 000 001
pico	p	10^{-12} =	0.000 000 000 001
femto	f	10^{-15} =	0.000 000 000 000 001
atto	a	10^{-18} =	0.000 000 000 000 000 001

Table 2.4: Most Common Prefixes Used in Introductory Physics

Mass	kg (kilogram = 10^{3} grams)
Length	mm (millimeter = 10^{-3} m), cm (centimeter = 10^{-2} m), km (kilometer = 10^{3} m), μm (micrometer = 10^{-6} m), nm (nanometer = 10^{-9} meter)
Time	ms (millisecond = 10^{-3} s), μs (microsecond, 10^{-6} s)
Power	MW (megawatt = 10^{6} W)
Frequency	kHz (kilohertz = 10^{3} Hz), MHz (megahertz = 10^{6} Hz)
Electric charge	μC (microcoulomb = 10^{-6} C)
Capacitance	μF (microfarad = 10^{-6} F), pF (picofarad = 10^{-12} F)
Electric Current	mA (milliampere = 10^{-3} A), μA (microampere = 10^{-6} A)
Resistance	kΩ (kilo ohm = 10^{3} Ω), MΩ (mega ohm = 10^{6} Ω)

Example 2.3.1
Write the following as decimal numbers with standard units:
a) 81 micrometers b) 32 picoseconds c) 11.2 femtometers d) 2.5 gigahertz

a) 81×10^{-6} m = 0.000081 m
b) 32 picoseconds = 32×10^{-12} s = 0.000000000032 s
c) 11.2 femtometers = 11.2×10^{-15} m = 0.0000000000000112 m
d) 2.5 gigahertz = 2.5×10^{9} Hz = 2500000000 Hz

Exercise
4. Rewrite the following quantities using prefixes.
    Wavelength of helium-neon laser = 0.0000006328 m
    Size of a virus = 0.0000001 m
    The radius of the nucleus of a carbon $[^{14}_{6}C]$ atom = 0.0000000000000029 m
    Rest energy of an electron = 511000 electron volts (eV).

2.4   Unit Conversions

Units can be converted from one system of units to another by multiplying by the appropriate factors, called conversion factors. *Conversion factors* are simply equivalence statements expressed in the form of ratios. For example, 1 in. = 0.0254 m, where the ratio is 0.0254. Some important conversions are listed in Appendix A. In unit conversion, multiply or divide by the appropriate conversion factor so that the unwanted units cancel, thus leaving the desired unit. This is shown in the next example.

Example 2.4.1
The posted speed limit on a highway is 65 mi/h. Determine the speed limit in kilometers per hour (km/h) and in meters per second (m/s).

Speed = 65 mi/h

1 mi = 1.609 km   That is, $1 = \left(1.609 \dfrac{km}{mi}\right)$.

Therefore, to convert mi/h to km/hr multiply by $\left(1.609 \dfrac{km}{mi}\right)$.

Thus, $65 \dfrac{mi}{h} = \left(65 \dfrac{mi}{h}\right)\left(1.609 \dfrac{km}{mi}\right) = 105 \dfrac{km}{h}$.

Also, 1 mi = 1609 m and 1 h = 3600 s. That is, $1 = \left(1609 \dfrac{\text{m}}{\text{mi}}\right)$ and $1 = \left(\dfrac{1\text{h}}{3600\,\text{s}}\right)$.

Therefore, $65\,\dfrac{\text{mi}}{\text{h}} = \left(65\dfrac{\cancel{\text{mi}}}{\cancel{\text{h}}}\right)\left(1609\dfrac{\text{m}}{\cancel{\text{mi}}}\right)\left(\dfrac{1\cancel{\text{h}}}{3600\,\text{s}}\right) = 29\,\dfrac{\text{m}}{\text{s}}.$

Example 2.4.2

Atmospheric pressure is 14.7 lb/in^2. Express atmospheric pressure in newtons per square meters (N/m^2), using the conversions 1 in = 0.0254 m and 1 N = 0.225 lb.

1 N = 0.225 lb and 1 in = 0.0254 m.   That is, $1 = \left(\dfrac{1\,\text{N}}{0.225\,\text{lb}}\right)$ and $1 = \left(\dfrac{1\,\text{in}}{0.0254\,\text{m}}\right)$.

Therefore, atmospheric pressure $= 14.7\,\dfrac{\text{lb}}{\text{in}^2} = \left(14.7\dfrac{\text{lb}}{\text{in}^2}\right)\left(\dfrac{1\,\text{N}}{0.225\,\text{lb}}\right)\left(\dfrac{1\,\text{in}}{0.0254\,\text{m}}\right)^2$

$$= 1.01 \times 10^5\,\dfrac{\text{N}}{\text{m}^2}.$$

Exercise

5. The highest waterfall is Angel Falls, whose height is 979.0 m. Express its height in miles.

6. The age of the universe is believed to be 15 billion years. Express the result in seconds.

7. A *light year* is the distance traveled by light in one year. Light travels 3.0×10^8 m in 1 second. An *astronomical unit* (AU) is the average distance from the earth to the sun, which is 1.5×10^8 km. Express 1 light year in terms of AU.

2.5   Dimension in Physics

The *dimension* of a physical quantity represents the *type* of the quantity, regardless of its units. The dimensions of length, mass, and time are expressed as [L], [M], and [T], respectively. For example, speed for a moving object, which is defined as distance over time, has the dimension [L]/[T].

Any valid mathematical equation must be dimensionally consistent. In deriving formulas or solving problems, dimension checks help verify each step of the derivation or solution. Each term in the equation must have the same dimensions.

Take an example that illustrates how a dimension check helps. The rate at which a pendulum moves back and forth depends on the length of the pendulum and the local acceleration due to gravity. Suppose you do not remember whether the period (time to complete one oscillation) of a pendulum is $T = 2\pi\sqrt{l/g}$ or $2\pi\sqrt{g/l}$. Here, l is the length of the pendulum, whose dimension is [L], and g is called the acceleration due to gravity whose dimension is $[L]/[T^2]$.

The dimension of the right-hand side of the first formula is

$$\sqrt{\frac{[L]}{[L/T^2]}} = \sqrt{[T^2]} = [T].$$

The dimension of the right-hand side of the second formula is

$$\sqrt{\frac{[L/T^2]}{[L]}} = \sqrt{\frac{1}{[T^2]}} = \frac{1}{[T]}.$$

The period is a time and it must have the dimension [T]. Obviously, the first equation is correct.

Important Note: A dimension check is not a guarantee that a formula or an answer is correct, because the dimensionless factors (such as 2π, ½, etc.) do not show up in the dimension check.

Exercises

8.   The speed, v, of an object is given by the equation $v = At^2 + Bt$ where t represents time. Determine the dimensions of A and B.

9. Show that the equations i) $x = vt + \frac{1}{2}at^2$ and ii) $v^2 = v_0^2 + 2ax$ are dimensionally correct, where the dimensions of x, v, v_0, a, and t are [L], [L]/[T], [L]/[T], $[L]/[T^2]$ and [T], respectively.

Answers to Exercises in Chapter 2

1. a) 2.72×10^5

   b) 5.34×10^{-7}

   c) 3.24×10^4

   d) 1.28×10^{-4}

   e) 5.4×10^{-6}

2. a) 2.0×10^2 kg

   b) 2.00×10^2 kg

3. a) $8.8 \times 10^0 \, m^2$

   b) 6.49 m/s

   c) 9.8×10^4 m

4. 632. 8 nm, 0.1 μm, 29 fm, 0.511 MeV

5. 0.608 miles

6. 4.7×10^{17} s

7. 1 light year = 6.3×10^4 AU

8. $[L]/[T^3]$, $[L]/[T^2]$

Chapter 3
Algebra: Dealing with Numbers and Equations in Physics

Introduction

Mathematics is the study of numbers and the geometry of objects. It is a tool to study nature and its changes. Algebra is essentially a system of rules and procedures for dealing with numbers and logically exploring the relationships between variables using various symbols. In physics, we deal with numbers, and one of our major concerns is to describe interdependences among physical quantities with equations. Algebra is very useful to explore such interdependences. For example, when an apple falls from a tree, its vertical distance (y) of fall can be described by an equation $y = Ct^2$, where C is a constant that is related to the gravity of the earth and t represents time.

3.1   Symbols and Letters

Different symbols and letters are used in algebra and, in general, in any mathematics to mean or abbreviate different concepts or operations. Some of the symbols and letters used in algebra, together with their meanings, are listed in Table 3.1. Greek letters are often used in mathematics (including algebra) and physics. The letters are shown in Table 3.2.

Table 3.1: Symbols and Letters

SYMBOL	MEANING	SYMBOL	MEANING
$=$	Equal to	$\therefore$	Therefore
$\approx$	Approximately equal to	$\sum$	Sum
$\sim$	Same order as	$\prod$	Product
$>$	Greater than	$\sqrt{\ }$	Square root
$\geq$	Greater or equal to	$x!$	Factorial of x
$<$	Less than	$\neq$	Not equal to
$\leq$	Less or equal to	∞	Infinity
$>>$	Much greater than	$\propto$	Proportional to
$<<$	Much less than	$\pm$	Plus or minus
$\equiv$	Equivalent to	$\mp$	Minus or plus
$\Rightarrow$	Implies	Δx	Change in x
$/$	Divided by		

Table 3.2: Greek Letters

	Capital	Lowercase		Capital	Lowercase
Alpha	A	α	Nu	N	ν
Beta	B	β	Xi	$\underline{\Xi}$	
		ξ			
Gamma	Γ	γ	Omicron	O	o
Delta	Δ	δ	Pi	Π	π
Epsilon	E	ε	Rho	P	ρ
Zeta	Z	ζ	Sigma	Σ	σ
Eta	H	η	Tau	T	τ
Theta	Θ	θ	Upsilon	Y	υ
Iota	I	ι	Phi	Φ	ϕ
Kappa	K	κ	Chi	X	χ
Lambda	Λ	λ	Psi	Ψ	ψ

In physics, it is common to use Greek letters to denote physical quantities. Although there are no accepted standards for representing such physical quantities by Greek letters, most introductory physics texts use the following Greek letters to represent the physical quantities mentioned.

α	Angular acceleration, coefficient of linear expansion, temperature coefficient of resistance, alpha ray
β	Coefficient of volume expansion, sound intensity level, beta ray
γ	Ratio of two specific heats of gases, electromagnetic photon or gamma ray
Δ	Change in
λ	Wavelength of sound, light, or any wave, and radioactive decay constant
τ	Torque and lifetime of unstable particles
σ	Stefan-Boltzmann constant, surface charge density, electrical conductivity
ρ	Density and resistivity of a material
ν	Frequency of sound, light, or any wave (texts also use f for frequency)
θ	Angular displacement
ω	Angular velocity, angular frequency
Ω	Electrical resistance in ohms
ε	Permittivity of a medium
μ	Coefficient of friction, permeability of a medium, magnetic dipole moment
η	Viscosity, efficiency of an engine (texts also use e for efficiency)
ϕ	Phase angle
ψ	Wave function of a particle

Dealing with Numbers

3.2   Numbers, Fractions, and Basic Rules

A *fraction* is the ratio of two numbers, a and b, and is expressed as a/b or $\dfrac{a}{b}$ (where $b \neq 0$). The numbers a and b are called the *numerator* and the *denominator* of the fraction, respectively.

The numbers ..., –4, –3, –2, –1, 0, 1, 2, 3, 4, ... are known as *integers*. Numbers that can be expressed as a fraction with an integer in its numerator and a nonzero integer in its denominator are called *rational numbers*. A rational number can also be expressed in decimal notation. For example, 10½ is a rational number, which, in decimal notation, is 10.5. *Irrational numbers* are numbers that cannot be expressed as a ratio of two integers. Their decimal notations do not terminate or repeat.

Some common irrational numbers that appear in physics are shown in Table 3.3.

Table 3.3: Common Irrational Numbers

$\pi = 3.141592\ldots$	$e = 2.718282\ldots$	$\sqrt{2} = 1.414213\ldots$
$\sqrt{3} = 1.732050\ldots$	$1/\sqrt{2} = 0.707106\ldots$	$\sqrt{3}/2 = 0.866025\ldots$
$\log 2 = 0.301029\ldots$	$\ln 2 = 0.693147\ldots$	$\ln 10 = 2.302585\ldots$

Rational and irrational numbers together constitute what are called *real numbers*.

Addition, Subtraction, Multiplication, and Division of Real Numbers

The following properties of real numbers are useful when you add or multiply two or more real numbers, or add 0 to a real number, or multiply a real number by 1.

For all real numbers a, b, and c,

i) $(a + b)$, $(a - b)$, $a \cdot b$ and a/b $(b \neq 0)$ are real numbers (Closure properties).

ii) $a + b = b + a$ and $a \cdot b = b \cdot a$ (Commutative properties).

Example: $2 + 7 = 9 = 7 + 2$, $3 \cdot 6 = 18 = 6 \cdot 3$

iii) $(a + b) + c = a + (b + c)$, $(ab)c = a(bc) = b(ac)$

and *(ab/c) = a(b/c) = b(a/c)*   (Associative properties).

Example: $2 + 5 + 9 = (2 + 5) + 9 = 16 = 2 + 14 = 2 + (5 + 9)$

$2 \cdot 5 \cdot 4 = 2(5 \cdot 4) = 2 \cdot 20 = 40 = 10 \cdot 4 = (2 \cdot 5)4$

iv) $a(b + c) = ab + ac$ (Distributive property).

Example: $2(3+4) = 2 \cdot 7 = 14 = 6 + 8 = 2 \cdot 3 + 2 \cdot 4$

v) $a + (-a) = 0$ and $a \cdot (1/a) = 1$ (Inverse properties).

vi) $a + 0 = a$ and $a \cdot 1 = a$ (Identity properties).

Rules for Combining Signs

Factors can be positive or negative. *When multiplying or dividing numbers of like sign, the result is positive; otherwise the result is negative.* For example,

$(+a)(+b) = (-a)(-b) = +ab$, and $(+a)(-b) = (-a)(+b) = -ab$.

Addition, Subtraction, Multiplication, and Division of Fractions

Use the following rules when adding, subtracting, multiplying, and dividing two fractions.

Addition and Subtraction	$\dfrac{a}{b} \pm \dfrac{c}{d} = \dfrac{ad \pm bc}{bd}$
Multiplication	$\dfrac{a}{b} \cdot \dfrac{c}{d} = \dfrac{ac}{bd}$
Division	$\dfrac{\dfrac{a}{b}}{\dfrac{c}{d}} = \dfrac{ad}{bc}$

Note: *Be careful of mistakes when you add two fractions.* For example,

$$\frac{a}{b} \pm \frac{c}{d} \text{ is not equal to } \frac{a \pm c}{b \pm d}. \text{ Also,}$$

$$\frac{1}{a} \pm \frac{1}{b} \text{ is not equal to } \frac{1}{a \pm b}.$$

In fact, $\dfrac{1}{a} \pm \dfrac{1}{b} = \dfrac{b}{ab} \pm \dfrac{a}{ba} = \dfrac{b \pm a}{ab}$.

Important Note: Find a common denominator when you add two or more fractions. In the preceding example, the common denominator is ab, which is obtained by multiplying the numerator and denominator of the first and second fractions by b and a, respectively. The next example will illustrate an application of this rule in physics.

Example 3.2.1 (from electricity)

The ability of an electrical component to oppose the flow of electricity through it is called its *electrical resistance R*. The *equivalent electrical resistance* R_{eq} of two electric components (such as two electric lightbulbs) R_1 and R_2 connected in a parallel combination is given by the formula $1/R_{eq} = 1/R_1 + 1/R_2$. Find the equivalent resistance of two resistances $R_1 = 100.0$ ohm and $R_2 = 200.0$ ohm. [Ohm is the unit of electric resistance.]

We have
$$a = R_1 = 100.0 \text{ ohm}$$
$$b = R_2 = 200.0 \text{ ohm}$$

This problem uses the rule of addition $\quad \dfrac{1}{a} + \dfrac{1}{b} = \dfrac{b + a}{ab}$.

Applying this rule, we get

$$\frac{1}{R_{eq}} = \frac{1}{R_1} + \frac{1}{R_2} = \frac{R_1 + R_2}{R_1 \cdot R_2}$$

Substitute the values of R_1 and R_2
$$= \frac{100.0 \, \text{ohm} + 200.0 \, \text{ohm}}{100.0 \, \text{ohm} \cdot 200.0 \, \text{ohm}}.$$

Thus,
$$R_{eq} = \frac{100.0 \, \text{ohm} \cdot 200.0 \, \text{ohm}}{100.0 \, \text{ohm} + 200.0 \, \text{ohm}}.$$

$$= \frac{20000}{300} \, \text{ohm} = 66.7 \, \text{ohm}.$$

Exercise

1. The equivalent resistance of three electrical resistances is given by

$$\frac{1}{R_{eq}} = \frac{1}{R_1} + \frac{1}{R_2} + \frac{1}{R_3}.$$

Show that $R_{eq} = \dfrac{R_1 R_2 R_3}{R_2 R_3 + R_3 R_1 + R_1 R_2}.$

Division with Zero

$\frac{15}{3} = 5$, since $5 \cdot 3 = 15$. Thus, $\frac{0}{x} = 0$, since $0 \cdot x = 0$. But $\frac{x}{0}$ is undefined, since no number multiplied by 0 gives x. Also, since any number multiplied by 0 gives 0, $\frac{0}{0}$ is indeterminate.

Important Note for Division with Zero	
$\dfrac{x}{0}$	Undefined
$\dfrac{0}{0}$	Indeterminate
$\dfrac{0}{x}$	0, if $x \neq 0$

Percent

A *percent* is the numerator of a fraction whose denominator is always 100. In other words, percent means *per hundred*. Thus, $20\% = \dfrac{20}{100} = 0.20$.

In problems involving percentages, the word "of" usually indicates multiplication. For example,

$$20\% \text{ of } 6000 = \dfrac{20}{100} \text{ x } 6000 = 1200.$$

In this example, 20% of 6000 = 1200, the percent is called the *rate* (r), 6000 is called the *base* (b) and 1200 is called the *percentage* (p). For any percentage problem,

$$rate \cdot base = percentage \text{ or } \quad r \cdot b = p.$$

Example 3.2.2

After examining 200 shirts, a quality control inspector found 10 with defective stitching, 6 with mismatched designs, and 1 with an incorrect label. What percent was defective?

Defective shirts = 10 + 6 + 2 = 18. Here, $b = 200$ and $p = 18$.

Use the relation $r \cdot b = p$

Substitute the values $r \cdot 200 = 18$

Divide by 200 to solve for r $r = \dfrac{18}{200} = 0.09$

Multiply by 100 to get the percentage $r = 0.09 \text{ x } 100\% = 9\%.$

3.3   The Number Line and Absolute Value

The number line shown below represents real numbers. Numbers to the *right* of 0 are called *positive* and numbers to the *left* of 0 are called *negative*. The number zero is neither positive nor negative.

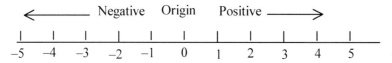

The magnitude of the distance between a number x and 0 is called the *absolute value* of x and is written as $|x|$. For example, since the distance between 4 and 0 is 4 units and the distance between –4 and 0 is also 4 units, $|4| = 4$ and $|-4| = 4$. Obviously, the absolute value of any number is either positive or zero.

The absolute value of a quantity is needed in physics, where the magnitude of the quantity is important, but not its sign. For example, when calculating the average deviation from the average result of an experimental variable, first calculate the absolute deviation of individual results $|x_1 - \bar{x}|$, $|x_2 - \bar{x}|$, ... (where x_1, x_2, ... are the results of individual measurements and $\bar{x}$ is the average value of the measurements). Then take the average by dividing the absolute deviations by the number of measurements.

Thus, if there are are N individual measurements, then

$$\text{Average deviation} = \frac{|x_1 - \bar{x}| + |x_2 - \bar{x}| + ... + |x_N - \bar{x}|}{N}$$

Example 3.3.1

a) Determine the absolute value of 3, –8 and 0.

b) The individual results of length measurements in a physics experiment are 4.0 cm, 3.9 cm, 3.7 cm, 4.1 cm, and 3.8 cm. Determine the absolute deviations of the individual results from the mean. Determine the average deviation.

a) $|3| = 3$ $|-8| = 8$ $|0| = 0$

b) The mean = (4.0 cm + 3.9 cm + 3.7 cm + 4.1 cm + 3.8 cm)/5 = 3.9 cm.

The absolute deviations of the individual results are as follows:

$|4.0\,\text{cm} - 3.9\,\text{cm}| = |0.1\,\text{cm}| = 0.1\text{cm},$ $|3.9\,\text{cm} - 3.9\,\text{cm}| = |0.0\,\text{cm}| = 0.0\text{cm},$

$|3.7\,\text{cm} - 3.9\,\text{cm}| = |-0.2\,\text{cm}| = 0.2\text{cm},$ $|4.1\,\text{cm} - 3.9\,\text{cm}| = |0.2\,\text{cm}| = 0.2\text{cm},$

$|3.8\,\text{cm} - 3.9\,\text{cm}| = |-0.1\,\text{cm}| = 0.1\text{cm}.$

Average deviation = (0.1 cm + 0.0 cm + 0.2 cm + 0.2 cm + 0.1 cm)/5 = 0.12 cm.

Proportion, Ratio, and Relationship between Quantities

3.4   Proportion and Ratio

The essence of physics is to describe and verify the relationships among physical quantities. The relationships are often simple: Two quantities may be directly proportional to each other, inversely proportional to each other, or inversely proportional to the square of the other.

Two quantities are said to be *directly proportional* to one another if an increase (or decrease) of the first quantity causes an increase (or decrease) of the second quantity by the same factor. If y is directly proportional to x, the direct proportionality is written as

$$y \propto x$$

The ratio y/x is a constant, say, k. That is, $y_1/x_1 = y_2/x_2 = k$.

> *Thus, in a direct proportion, the relationship is described by an equation, $y = kx$.*

Look at the following data for the diameter and the circumference of a circle.

Diameter, d (m)	Circumference, C (m)
0.240	0.754
0.420	1.319
0.720	2.262

The data show that the ratio of the circumference C to the diameter d is always 3.14, which is called π. We can say that the circumference of a circle is directly proportional to its diameter, and $C = \pi d$, where π is the constant of proportionality.

In general, in a direct proportion, $\dfrac{a}{b} = \dfrac{c}{d}$, a and d are called the *extremes* and b and c are called the *means*. Multiplying both sides by bd, we find that

$$b d \cdot \frac{a}{b} = b d \cdot \frac{c}{d} \qquad \text{or} \qquad a \cdot d = b \cdot c.$$

That is, the product of the extremes is equal to the product of the means.   This fact is useful in mathematics in solving for an unknown variable, say, x, from an equation, such as

$$\frac{x}{4} = \frac{21}{7}.$$

Thus, $7x = 4 \cdot 21$ or $x = \frac{4 \times 21}{7} = 12.$

Example 3.4.1

If y is directly proportional to x, and $x = 2$ when $y = 16$, what is y when $x = 6$?

Since y is directly proportional to x, $y_1/x_1 = y_2/x_2 = k$.

Substitute the values $\frac{y}{6} = \frac{16}{2}$

Multiply by $2 \cdot 6$ $2y = 16 \cdot 6 = 96$

Divide by 2 $y = \frac{96}{2} = 48$

As mentioned, in physics you will encounter many situations where the quantities are related in direct proportions. A simple example is the stretching of an ordinary helical spring. The spring has a certain length to begin with, and when it is suspended vertically with weights attached to the bottom, the length of the spring increases. If the amount of stretch is not too long, the amount of weight F (measured in pounds or newtons) and the amount of stretch x are directly proportional to each other [Figure 3.1]. This is known as *Hooke's law*.

Thus, $F = k\,x$.
The constant k is called the *stiffness constant* or *spring constant* of the spring.

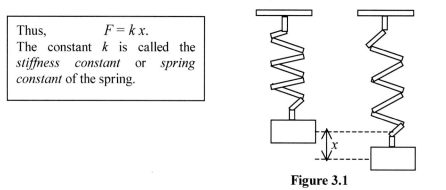

Figure 3.1

Example 3.4.2 (from mechanics)

A spring is suspended vertically from a fixed support (Figure 3.1). The stiffness constant of the spring is $k = 10.0$ N/m. When a weight is attached to the bottom of the spring, the spring stretches by 0.1 m. Determine the weight attached to the spring. The weight will be in newtons (N).

List data, check units and convert the unit(s) if necessary. We have
$$k = 10.0 \text{ N/m}$$

and $x = 0.1 \text{ m}.$

The units are correct; no conversion is necessary. The sketch of the problem is shown in Figure 3.1 where $x = 0.1$ m. The applied weight and the amount of stretch are related in direct proportions, as expressed by Hooke's law, $F = k\,x$. Apply Hooke's law to solve for the weight:

$$F = k\,x \;\; = (10.0 \text{ N/m}) \cdot (0.1 \text{ m}) = 1.0 \text{ N}.$$

Example 3.4.3 (from thermal physics)

Charles' law is an important law governing the behavior of gases; it states that if the pressure of a gas is kept constant, its volume (V) is *directly proportional* to its absolute temperature (T) measured in kelvins (K). Use this relationship to determine the volume of 2 l of a gas kept in an open container when it heats up from 77 K to room temperature 300 K.

Initial volume $V_1 = 2$ l
Initial temperature, $T_1 = 77$ K
Final temperature, $T_2 = 300$ K
Final volume, $V_2 = ?$

Since the container is open, pressure is always constant. Volume and temperature are related in direct proportions, since pressure is constant. Thus, the ratio V/T is constant, or

$$\frac{V_1}{T_1} = \frac{V_2}{T_2}.$$

Multiply both sides by T_2 to solve for V_2

$$V_2 = \frac{V_1}{T_1} T_2$$

Substitute the values of V_1, T_1, and T_2

$$V_2 = \left(\frac{21}{77\,K}\right) 300\,K = 8\,l.$$

As temperature has increased, the final volume has increased, so the result makes sense.

Exercise

2. Hubble's law: According to Hubble's law, the speed of recession (v) of a galaxy is *directly proportional* to its distance (d) from the earth and is given by $v = Hd$, where the proportionality constant H is called Hubble's constant, which is 0.022 in the unit of m/(s·light-year) (That is, the constant is 0.022 when speed is measured in the unit of *meter per seconds*, and the distance is measured in *light-years*.) A galaxy is found to be moving away from the earth at a speed of 5.0 X 10^5 m/s. Determine the distance of the galaxy in light years.

3. A hand exerciser uses a coiled spring that follows Hooke's law, which states that the force needed to compress the spring is *directly proportional* to the amount of compression. If a force of 90 N is needed to compress the spring by 0.02 m, determine the force required to compress the spring by 0.047 m. (*Hint*: First find the constant of proportionality by using Hooke's law, $F = kx$.)

4. When the volume of a closed container is constant, the pressure P of gas inside the container is *directly proportional* to its temperature T in kelvins (K). An automobile tire is filled to a pressure of 2.1 X 10^5 Pascal (Pa) at a temperature of 283 K. After the car is driven for 1 hour, the temperature within the tire rises to 310 K. What is the final pressure of the gas?

5. The pressure of water inside a lake is directly proportional to its depth. If the pressure of water at a depth of 1.0 m is 9.8 X 10^3 N/m^2, what is the pressure of water at a depth of 18.0 m?

Graph of Direct Proportionality Relationship

In a Cartesian graph, named after the seventeenth-century mathematician René Descartes, a rectangular coordinate system is formed by two perpendicular lines. The horizontal line is called the x-axis and the vertical line is called the y-axis. The point where the axes cross is called the *origin*.

When a quantity y has a direct proportionality with another quantity x, the graph of y versus x is always a straight line passing through the origin, as shown. The graph $y = kx$ is shown in Figure 3.2. In the graph, the change of the quantity x is labeled as Δx (which is often called "run") and the corresponding change in y is labeled as Δy (which is often called "rise").

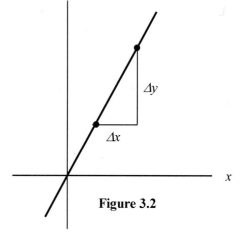

Figure 3.2

Here,

$$\Delta x = x_2 - x_1$$

$$\Delta y = y_2 - y_1$$

where (x_1, y_1) and (x_2, y_2) are coordinates of the two points on the line. The constant of proportionality between Δy and Δx is also k. Thus,

$$\Delta y = k\, \Delta x.$$

The steepness of the line is measured by the ratio $\Delta y / \Delta x$ and is called the *slope* of the line. Thus,

$$\text{slope} = \frac{y_2 - y_1}{x_2 - x_1} = \frac{\Delta y}{\Delta x} = k.$$

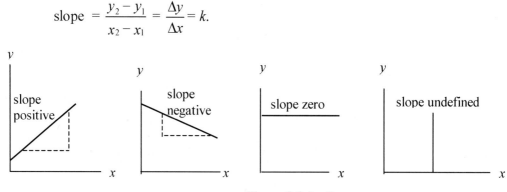

Figure 3.3 (a-d)

Therefore, in the direct proportionality between two quantities, the proportionality constant is the slope of the line representing the relationship between the quantities. The slope of a line can be *positive, negative, zero* or *undefined*, as shown in Figure 3.3 (a–d).

Note:
Slope *positive* means *y increases as x increases.*
Slope *negative* means *y decreases as x increases.*
Slope *zero* means *y* does not change—that is, the line is *parallel to the x-axis.*
Slope *undefined* means *x* does not change—that is, the line is *parallel to the y-axis.*

3.5 Simple Inverse Proportion

In physics, you will encounter situations where one physical quantity increases and the other quantity decreases in such a way that their product always stays the same. In this case, they are said to be in *inverse proportion*. In inverse proportion, when one quantity approaches zero, the other quantity becomes extremely large, so that the product still remains the same. Mathematically, if *y* is in inverse proportion with *x*,

$$y \propto \frac{1}{x} \qquad \text{or} \qquad y = \frac{k}{x} \quad \text{where } k \text{ is the proportionality constant.}$$

Thus, when *y* is inversely proportional to *x*, $xy = k$.

In other words, when *x* changes from x_1 to x_2, *y* changes y_1 to y_2 so that

$$x_1 y_1 = x_2 y_2 = k.$$

This type of behavior is illustrated Figure 3.4 (for an arbitrary choice of $k = 10$).

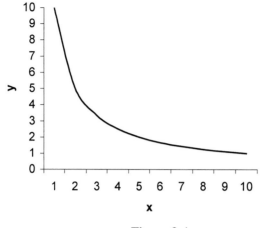

Figure 3.4

Example 3.5.1 (from thermal physics)

Boyle's law is an important law governing the behavior of ideal gases. According to the law, if the temperature of a gas is kept constant, its pressure, P, is *inversely proportional* to its volume, V. A cylindrical flask is fitted with an airtight piston. Contained within the flask is an ideal gas. Initially, the pressure applied by the piston is 12×10^4 Pa and the volume of the gas is 7.5×10^{-3} m^3. Assuming that the system is always at the temperature of 290 K, determine the volume of the gas when its pressure increases to 18×10^4 Pa. We have

$$P_1 = 12 \times 10^4 \text{ Pa} \qquad\qquad P_2 = 18 \times 10^4 \text{ Pa}$$
$$V_1 = 7.5 \times 10^{-3} \text{ m}^3 \qquad\qquad V_2 = ?$$

Following is a sketch of the problem:

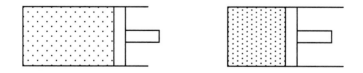

Since the pressure P is inversely proportional to the volume V according to Boyle's law, the product PV remains constant.

That is,
$$P_1 V_1 = P_2 V_2$$

Divide by P_2 to solve for V_2

$$V_2 = \frac{P_1 V_1}{P_2}$$

Substitute the values of P_1, V_1 and P_2

$$= \frac{\left(12 \times 10^4 \, \text{Pa}\right) \times \left(7.5 \times 10^{-3} \, \text{m}^3\right)}{18 \times 10^4 \, \text{Pa}}$$

$$= \frac{12 \times 7.5}{18} \times 10^{-3} \, \text{m}^3$$

$$= 5 \times 10^{-3} \, \text{m}^3$$

Since the pressure increased, the final volume had decreased, as expected.

Exercise

6. According to Kepler's second law of planetary motion, the speed v of a planet around the sun is *inversely proportional* to the distance r of the planet from the sun. If the speed of a planet is 5.0×10^4 m/s when it is at its shortest distance of 1.2×10^{15} m from the sun, find its speed when it is at its longest distance of 2.2×10^{15} m from the sun.

3.6 Inverse Square Proportion

Inverse square dependence with distance is found in different laws of nature. In physics, you will encounter situations where a physical quantity decreases uniformly in all directions with the square of the distance. For example, sound intensity (also light intensity), which is a measure of the sound power (or light power) distributed over unit area, decreases as the inverse square of the distance from the source of sound. The force of gravitation due to a spherical mass (such as a planet or a star) decreases as the inverse square of the distance from the center of the mass. The electrostatic force due to a point electric charge decreases as the square of the distance from the charge.

Mathematically, if y varies inversely with the square of x, then

$$y \propto \frac{1}{x^2} \qquad \text{or} \qquad y = \frac{k}{x^2},$$

where k is the proportionality constant.

This relationship is illustrated in Figure 3.5 (for an arbitrary choice of $k = 10$).

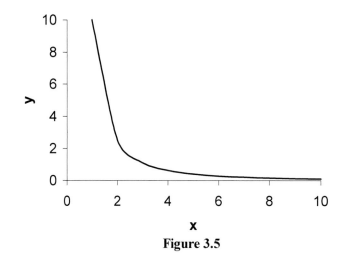

Figure 3.5

Example 3.6.1 (from sound)

A small source of sound emits sound equally in all directions. The intensity of sound is given by the equation $I = k/r^2$, where k is a constant and r is the distance from the source. If the sound intensity is 0.01 W/m^2 at a distance 1 m from the source, find the intensity at a distance of 10 m from the source.

Initial distance, $r_1 = 1$ m

Sound intensity, $I_1 = (0.01 \text{ W/m}^2)$

Final distance, $r_2 = 10$ m

Sound intensity, $I_2 = ?$

See the accompanying sketch of the problem.
The intensity of sound is given by the equation $I = k/r^2$. Use this equation to solve the problem. However, in this problem, you do not need to know the value of k. From the inverse square relationship, we have

$$I_1 = \frac{k}{r_1^2} \qquad \text{and}$$

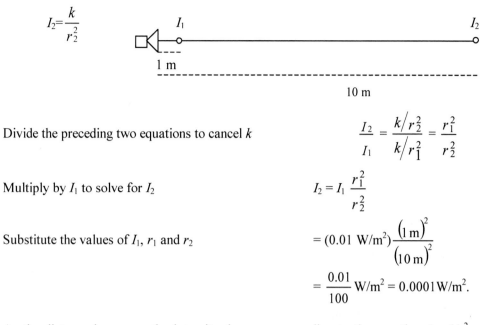

$$I_2 = \frac{k}{r_2^2}$$

I_1 I_2

1 m

10 m

Divide the preceding two equations to cancel k
$$\frac{I_2}{I_1} = \frac{k/r_2^2}{k/r_1^2} = \frac{r_1^2}{r_2^2}$$

Multiply by I_1 to solve for I_2
$$I_2 = I_1 \frac{r_1^2}{r_2^2}$$

Substitute the values of I_1, r_1 and r_2
$$= (0.01 \text{ W/m}^2)\frac{(1\text{ m})^2}{(10\text{ m})^2}$$

$$= \frac{0.01}{100} \text{ W/m}^2 = 0.0001\text{ W/m}^2.$$

As the distance increases, the intensity decreases according to the equation $I = k/r^2$. That is, the value of the sound intensity at 10 m should be less that the value at 1 m. Thus, the result makes sense.

Exercises

7. A lamp emits light uniformly in all directions, and the intensity is given by an inverse square law, $I = k/r^2$. If the light intensity reaching an object at a distance of 0.20 m is 100 lux, determine the light intensity at 1.5 m from the lamp.

8. The force of gravity exerted by the earth on an object varies inversely as a square of the distance from the center of the earth. Determine the ratio of the force of gravity on an astronaut when he is on the surface of the earth and at a height of 10,000 km from the surface of the earth. The radius of the earth is 6.38 X 10^6 m.

Rules of Algebra

3.7   Power and Exponents

You will find that many equations in physics are nonlinear. For example, the relation between distance and time for a freely falling object (such as a falling apple from a tree) is nonlinear. In the expression x^m, x is called the *base* and m is called the *exponent.* In a nonlinear equation, one or more variables have exponents or powers.

$$x^m = x \cdot x \cdot x \cdot x \cdot x \cdot ... \cdot \text{(product of } m \text{ factors of } x)$$

Example 3.7.1 (from heat transfer)

Stefan's law: Energy radiation from hot bodies varies as the *fourth power* of its absolute temperature, $E = \sigma T^4$, where σ is a constant. Determine by which factor the energy radiated from the sun will be reduced, if its surface temperature changes from 6000 K to 5600 K.

For this problem, $T_1 = 6000$ K and $T_2 = 5600$ K,

Radiant energy will be E_1 and E_2 when the temperatures are T_1 and T_2, respectively. Use Stefan's law.

$T = 6000$ K $T = 5600$ K

From Stefan's law,

$$E_1 = \sigma T_1^4$$
$$E_2 = \sigma T_2^4.$$

Divide to cancel the constant σ

$$E_1/E_2 = \sigma T_1^4/\sigma T_2^4 = T_1^4/T_2^4.$$

Substitute the values of T_1 and T_2

$$E_1/E_2 = (6000)^4/(5600)^4$$

$$= 1.3.$$

Thus, the energy radiated from the sun will be reduced by the factor of 1.3.
Since the temperature has decreased, the radiated energy has decreased, as expected from Stefan's law.

3.8   Rules for Exponents

Use the following rules for exponents where m and n are natural numbers:

Product rule	$x^m \cdot x^n = x^{m+n}$
Quotient rule	$\dfrac{x^m}{x^n} = x^{m-n} \quad (x \neq 0)$
First power rule	$(x^m)^n = x^{m \cdot n}$
More power rules	$(xy)^m = x^m \cdot y^m$
	$\left(\dfrac{x}{y}\right)^m = \dfrac{x^m}{y^m}$
Zero exponent rule	$x^0 = 1$ (where x is a nonzero real number; 0^0 is undefined)
Negative exponent rule	$x^{-m} = \dfrac{1}{x^m}$
Roots	$x^{1/m} = \sqrt[m]{x}$

The preceding rules can be easily verified. For example,

$10^3 \times 10^2 = (10 \times 10 \times 10) \times (10 \times 10) = 10^5 = 10^{(3+2)}$. This verifies the product rule. Similarly,

$$\frac{10^3}{10^2} = \frac{10 \times 10 \times 10}{10 \times 10} = 10 = 10^1 = 10^{(3-2)}.$$

This verifies the quotient rule. Also,

$$(10^2)^3 = (10 \times 10)^3 = (100)^3 = 100 \times 100 \times 100 = 100\ 00\ 00 = 10^6 = 10^{3 \cdot 2}.$$

This verifies the first power rule.

Note that the zero exponent rule follows from the quotient rule, since

$$1 = \frac{x^2}{x^2} = x^{2-2} = x^0.$$

The negative exponent rule also follows from the zero exponent rule and the quotient rule, because

$$\frac{1}{x^m} = \frac{x^0}{x^m} = x^{0-m} = x^{-m}.$$

Example 3.8.1
Simplify each expression:

i) $(-3x^4y^2)^2$

ii) $\left(\dfrac{4x^2}{2y^3}\right)^2$

iii) $\dfrac{\left(x^3y^4\right)^2}{\left(xy\right)^4}$

iv) $\left(-\dfrac{x^2y^4}{xy^{-2}}\right)^{-3}$

i) $(-3x^4y^2)^2 = (-3)^2(x^4)^2(y^2)^2 = 9x^{4\cdot2}y^{2\cdot2} = 9x^8y^4$

ii) $\left(\dfrac{4x^2}{2y^3}\right)^2 = \dfrac{4^2 x^{2\cdot2}}{2^2 y^{3\cdot2}} = \dfrac{16x^4}{4y^6} = \dfrac{4x^4}{y^6}$

iii) $\dfrac{\left(x^3y^4\right)^2}{\left(xy\right)^4} = \dfrac{\left(x^3\right)^2\left(y^4\right)^2}{x^4y^4} = \dfrac{x^{3\cdot2}y^{4\cdot2}}{x^4y^4} = \dfrac{x^6y^8}{x^4y^4} = x^{6-4}y^{8-4} = x^2y^4$

iv) $\left(-\dfrac{x^2y^4}{xy^{-2}}\right)^{-3} = \left(-x^{2-1}y^{4-(-2)}\right)^{-3} = \left(-x^1y^6\right)^{-3} = -x^{-3}y^{-18}$

Example 3.8.2 (from acoustics)
In a chromatic music scale, there are 12 equal music intervals, and the frequency of every 13th note is double that of the beginning note. That is, the frequency of each note increases by a factor of $2^{1/12}$ from the preceding note. If the frequency of the

middle C is 261.63 Hz ("Hz" abbreviates hertz, the unit of frequency), what is the frequency of the next note in the music scale? [The frequency of a note tells how many cycles of vibration the source of sound completes in one second as it produces sound.]

Frequency of the middle C = 261.63 Hz.

Since the frequency of each note is increased by the factor $2^{1/12}$, to get the frequency of the next note, multiply that frequency by $2^{1/12}$.

Thus, frequency of the next note of middle C

$$= 261.63 \text{ Hz} \times 2^{1/12}$$
$$= 261.63 \text{ Hz} \times 2^{0.0833}$$

Use a calculator to find $2^{0.0833}$
$$= 261.63 \times 1.0595$$
$$= 277.20 \text{ Hz}$$

Exercise
Simplify each expression

9. $(x^2y^4)^2(x^3z^2)^{-2}(z^3x^2)^{-1}$

10. $\left(\dfrac{x^2y^4}{xy^2}\right)^0$

11. $\dfrac{\left(xy^{-3}z\right)^3}{\left(x^{-2}y\right)^2}$

12. $\left(\dfrac{2a^2b^4c^{-2}}{8a^{-1}b^2c^{-4}}\right)^{-3}$

3.9   Factoring a Polynomial and an Equation

An algebraic expression containing the sum of one or more terms that contain whole number exponents is called a *polynomial*. A polynomial that has only one term is called a *monomial*, whereas polynomials with two and three terms are called *binomials* and *trinomials*, respectively. For example, $2x + 4$ is a binomial and $5x^2 + 2x + 3$ is a trinomial. The *degree* of the monomial ax^n ($a \neq 0$) is n. The degree of a monomial with more than one variable is the sum of the exponents of the variables. For example, the degree of the monomial $2x^2y^4$ is $(4 + 2)$ or 6.

An *equation* is an equivalent statement that indicates that the two quantities are equal. For example, $x - 2 = 4$ is an equation. Here, x is the variable of the equation. $x - 2$ is the left-hand side, and 4 is the right-hand side, of the equation. All the values of

x that make the left-hand and right-hand sides equal are called solutions of the equation.   An equation such as $a \cdot x + b = 0$ $(a \neq 0)$ that involves first-degree polynomials is called a *linear equation*. In general, an equation of the form $ax^2 + bx + c = 0$ (where a, b, and c are real numbers and $a \neq 0$) is called a *quadratic equation*.

Common terms can be *factored* out from a polynomial or an equation to simplify it. A trinomial of the form $ax^2 + bx + c$, where $a \neq 0$ and a, b, and c are all integers, will factor into two binomials with integer coefficients if the value of $b^2 - 4ac$ is a perfect square. By direct multiplication,

$$(x + p)(x + q) = x^2 + px + qx + pq = x^2 + (p + q)x + pq$$

Thus, the coefficient of the middle term is the sum of p and q and the last term is the product of p and q. Note that, in this case, $b^2 - 4ac = (p + q)^2 - 4pq = (p - q)^2$.

Example 3.9.1
Factor $x^2 - 2x - 63$.

Compare the given trinomial with $ax^2 + bx + c$.

In this case, $\qquad\qquad\qquad\qquad a = 1, b = -2$ and $c = -63$

$$b^2 - 4ac = (-2)^2 - 4 \cdot 1 \cdot (-63) = 4 + 252 = 256 = 16^2$$

Since $b^2 - 4ac$ is a perfect square, the given trinomial will factor into two binomials.

Now, try to write the coefficient (-2) as $(p + q)$ and the last term (-63) as pq.

Product of the factors	Sum of the factors		
−63 (1)	−63 + 1	=	−62
63 (−1)	63 − 1	=	62
7 (−9)	7 − 9	=	−2
−7 (9)	−7 + 9	=	2

Obviously, the choice is $p = 7$ and $q = -9$, since, in this case, $p + q$ is –2 and pq is –63.

Thus, $\qquad\qquad\qquad\qquad x^2 - 2x - 6 = x^2 + 7x - 9x + 7(-9)$

$$= x(x + 7) - 9(x + 7)$$

$$= (x + 7)(x - 9).$$

Remember the following formulas for factoring.

$$ax + ay + az = a(x + y + z)$$

$$(x + y)^3 = x^3 + 3x^2y + 3xy^2 + y^3$$

$$x^2 - y^2 = (x + y)(x - y)$$

$$(x - y)^3 = x^3 - 3x^2y + 3xy^2 - y^3$$

$$(x + y)^2 = x^2 + 2xy + y^2$$

$$x^3 + y^3 = (x + y)(x^2 - xy + y^2)$$

$$(x - y)^2 = x^2 - 2xy + y^2$$

$$x^3 - y^3 = (x - y)(x^2 + xy + y^2)$$

All of these formulas can be easily verified by direct multiplication.

Exercise

13. Factor the following trinomials:

a) $t^2 - 7t + 12$ b) $2x^2 - 4x - 48$ c) $96a + 6ab^2 + 48ab$

d) $2a^2 + 5a + 3$ (*Hint*: The first terms of the binomial factors are $2a$ and a.)

Solving Equations

3.10   Properties of an Equation

To solve an equation, remember the following *properties*.

If $x = y$ represents an equation, and a is a real number, then

$$x \pm a = y \pm a$$

$$x \cdot a = y \cdot a$$

$$\frac{x}{a} = \frac{y}{a} \qquad \text{if } a \neq 0$$

$$x^n = y^n$$

$$\sqrt[n]{x} = \sqrt[n]{y}$$

Note: *Any mathematical operation that is applied to one side of an equation must also be applied to the other side.*

Important Note: If any side of an equation contains more than one term, the operation must be applied to the *entire side*. That is, if $c = a + b$,

$$3c = 3(a + b) \qquad \qquad (\text{not } 3a + b)$$

$$\frac{c}{2} = \frac{a}{2} + \frac{b}{2} = \frac{a + b}{2} \qquad (\text{not } \frac{a}{2} + b)$$

$$c^2 = (a + b)^2 \qquad \qquad (\text{not } a^2 + b^2)$$

$$\sqrt{c} = \sqrt{a + b} \qquad \qquad (\text{not equal to } \sqrt{a} + \sqrt{b})$$

Example 3.10.1 (from electricity)
The equation $V = IR$ is called Ohm's law in electricity, where V, I, and R are voltage, current, and resistance, respectively. Solve the equation for the current I when $V = 120$ volts and $R = 10$ ohms. The unit of current is ampere.

For this problem, $V = 120$ volts $R = 10$ ohm

Use Ohm's law to solve the problem.

$$V = IR$$

Divide both sides by R to solve for I $\dfrac{V}{R} = \dfrac{IR}{R} = I$

Substitute the values of V and R $I = \dfrac{V}{R} = \dfrac{120 \text{ volts}}{10 \text{ ohms}} = 12$ amperes.

3.11   Linear Equations Involving One Variable

An equation of the form $ax + b = 0$ $(a \neq 0)$ is called a linear equation involving the unknown variable x. The basic operations of an equation (mentioned in the previous section) are sufficient to solve all linear equations. The objective is to select the proper operations to determine the unknown quantity x.

Example 3.11.1
Solve $3x + 2 = 5x - 1$.

$$3x + 2 = 5x - 1$$

Add –2 on both sides $3x + 2 - 2 = 5x - 1 - 2$

$$3x = 5x - 3$$

Add –5x on both sides so that right hand
side does not contain any term containing x $3x - 5x = 5x - 3 - 5x$

$$-2x = -3$$

Divide both sides by –2 $\dfrac{-2x}{-2} = \dfrac{-3}{-2}$

$$x = \frac{3}{2}$$

To verify your answer, substitute $x = 3/2$ in the equation. The left-hand side is then 3 $(3/2) + 2 = 9/2 + 2 = 13/2$. The right-hand side is 5 $(3/2) - 1 = 15/2 - 1 = 13/2$, as expected.

Exercises
Solve the following equations for x.

14. $4x + 8 = 0$ 15. $4x + 5 = 9x - 20$ 16. $\dfrac{3}{x-2} = \dfrac{4}{x-1}$

Example 3.11.2

If $F = kqQ/r^2$; solve for k.

Multiply by r^2 $Fr^2 = kqQ$

Divide by qQ to cancel qQ from the right $\dfrac{F r^2}{qQ} = \dfrac{kqQ}{qQ} = k$

Therefore, $k = \dfrac{F r^2}{qQ}$.

Exercises
Solve the following equations for the unknown in terms of other quantities in the equations.

17. $F = GMm/r^2$; solve for G.   18. $(P + a/V^2)(V - b) = RT$; solve for P.

19. $L = L_0(1 + \alpha\, \Delta T)$; solve for α 20. $1/d_0 + 1/d_i = 1/f$; solve for f.

3.12   Linear Equations Involving Two Variables

An equation of the form $ax + by + c = 0$ or of the form $y = mx + b$ is called a *linear equation* in x and y. Linear relationships between two physical variables are common. For example, when a moving object speeds up at a constant rate a, its velocity can be expressed by an equation $v = v_0 + at$ where v_0 is the initial velocity of the object and t represents time. When a solid is heated, its length, L, increases with temperature, T, and the relationship can be approximated by a linear equation $L = L_0 + \alpha(T - T_0)$, where L_0 and T_0 are the initial length and temperature, respectively, of the solid and α is a constant for the solid. There are many other situations when the relationship between two physical variables can be approximated as linear.

Example 3.12.1 (from kinematics)

When a moving object speeds up at a constant rate a (acceleration) starting from an initial velocity v_0, its final velocity (v) increases linearly with time (t) and is given by $v = v_0 + at$. An object starts with an initial velocity of 1.1 m/s and speeds up at a constant rate (called acceleration) of 2.0 m/s^2. What is the speed of the object after 10.0 s?

For this problem, $v_0 = 1.1$ m/s,

$a = 2.0$ m/s^2,

$t = 10.0$ s, and

$v = ?$

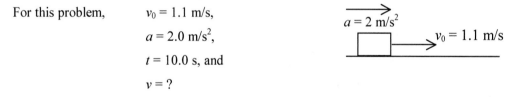

The sketch of the problem is shown at the right. Use the relationship $v = v_0 + at$ to solve for v.

$$v = v_0 + at.$$

Substitute the values of v_0, a, and t $v = 1.1$ m/s $+ (2.0$ m/s$^2) \times 10.0$ s

$= 21.1$ m/s

$= 21$ m/s (rounding to two significant

digits).

Since the object is accelerating, the final velocity is more than the initial velocity. So the result makes sense.

Exercises

21. The ability of a material to resist the flow of electricity is called its *electrical resistance R.* The electrical resistance of a metal increases linearly with temperature T according to the equation $R = R_0 + \alpha_0(T - T_0)$, where α_0 is a constant of the metal, called the *temperature coefficient of resistance.* Find the change in resistance, $R - R_0$, of platinum when the temperature change $(T - T_0)$ is 100 °C. α_0 for platinum is 0.003927 /C°.

22. The increase in length, ΔL, of brass of length L for a temperature rise by ΔT can be approximated by the linear equation $\Delta L = \alpha L \Delta T$.   Find the temperature change of a piece of brass of length 2.0 m when its length increases by 1.0 mm. The constant α for brass is 19.0×10^{-6} (in the unit of /C°).

3.13   Solving Two Simultaneous Linear Equations

Two linear equations can be solved by the *substitution method*, the *addition method*, or the *graphical method*. The method of determinant (Cramer's rule) can also be used to solve two linear equations, but this method is more useful to solve several linear equations each containing more than two variables.

Substitution Method

To solve two equations involving x and y by substitution

i)   Solve one equation for one variable (say, x) in terms of the other variable (say, y).

ii)   Substitute the resulting expression for the variable x into the other expression. Solve the new equation for y.

iii)   Substitute the known value of y into one of the original equations to solve for x.

iv)   Verify that x and y satisfy the original equations.

Example 3.13.1

Solve $\begin{cases} x - y = 2 \\ x + 2y = 8 \end{cases}$ by the method of substitution.

$$x - y = 2$$

Add $+ y$ on both sides	$x - y + y = 2 + y$
or	$x = 2 + y$

$$x + 2y = 8$$

Substitute $2 + y$ for x	$(2 + y) + 2y = 8$
or	$2 + 3y = 8$
Subtract 2 from both sides	$3y = 8 - 2$
or	$3y = 6$
Divide both sides by 3	$y = \dfrac{6}{3}$
	$= 2$

Substitute $y = 2$ into the original equation, $x - y = 2$

$$x - 2 = 2$$

Add $+2$ on both sides	$x = 4$

The solution is $\begin{cases} x = 4 \\ y = 2 \end{cases}$.

(To verify your answer, substitute $x = 4$ and $y = 2$ in the left hand side of both of the equations. $x - y = 4 - 2 = 2$ and $x + 2y = 4 + 2 \cdot 2 = 4 + 4 = 8$. Thus, x and y satisfy both the equations.)

Addition Method

To solve two equations by the method of addition:

i)   Write both equations in the form $ax + by = c$.

ii)   Multiply (if necessary) one or both of the equations by nonzero quantities so that the coefficients of x (or y) in the equations become equal, but have opposite signs.

iii)   Add the resulting equations to eliminate x (or y), and solve for y (or x).

v)   Substitute the known value of y (or x) into one of the original equations to solve for x (or y).

iv)   Verify that x and y satisfy the original equations.

Example 3.13.2

Solve $\begin{cases} 2x - y = 6 \\ x + 2y = 8 \end{cases}$ by the addition method.

Multiply the second equation by –2

$$2x - y = 6$$
$$-2x - 4y = -16$$

Add $-5y = -10$

Divide by –5 $y = \dfrac{-10}{-5}$

$$= 2$$

Substitute $y = -2$ into the original equation

$$2x - 2 = 6$$

Add 2 on both sides $2x = 6 + 2$

or $2x = 8$

Divide by 2 $x = 4$

The solution is $\begin{cases} x = 4 \\ y = 2 \end{cases}$.

Exercises

Solve the following equations by the substitution method and the addition method.

23. $x + y = 3$, $x - y = 1$

24. $\dfrac{x}{2} + \dfrac{y}{3} = 2$, $2x - y = 1$

25. $5x - 4y = 1$, $6y = 10x - 4$

26. $x = 2(4 - y)$, $3x = 6(8 - 3y)$

27. $2(x + 3) + 7(2 - y) = 4$, $2(x - 3) - 4(y - 3) = 2$

28. $\dfrac{x+3}{2} + \dfrac{y+2}{3} = 5$, $\dfrac{x-1}{2} + \dfrac{y-1}{3} = 2$

3.14   Solution of Quadratic Equations

An equation of the form $ax^2 + bx + c = 0$, where a, b, and c are real numbers and $a \neq 0$, is called a *quadratic equation*.

Sometimes you will encounter the simplest quadratic equation, such as $x^2 = a$. To solve this equation you need to take the square root of both sides:

$$\sqrt{x^2} = \sqrt{a}$$

$$x = \pm\sqrt{a} \ .$$

The symbol $\pm$ means that $\sqrt{a}$ carries either a positive sign or a negative sign. This should be obvious since the square of both $+\sqrt{a}$ and $-\sqrt{a}$ give a. Usually, in a physical problem, the situation will determine which sign or signs are appropriate.

Important Note:

$$\sqrt{ab} = \sqrt{a} \times \sqrt{b}$$

But $\sqrt{a+b}$ is not equal to $\sqrt{a} + \sqrt{b}$.

Example 3.14.1

Solve $4x^2 - 2 = 2x^2 + 16$.

To solve this equation, rearrange the terms so that left-hand side contains only the term in x^2. To cancel 2 from the left-hand side, add 2 on both sides. Also, add $-2x^2$ on both sides to cancel $2x^2$ from the right. Thus, add $(2 - 2x^2)$ on both sides.

$$4x^2 - 2 = 2x^2 + 16.$$

Add $(2 - 2x^2)$ $\qquad$ $4x^2 - 2 + (2 - 2x^2) = (2 - 2x^2) + 2x^2 + 16$

$\qquad$ or $\qquad$ $4x^2 - 2x^2 = 2 + 16$

$$2x^2 = 18$$

Divide boh sides by 2 $\qquad$ $x^2 = 9$

Take the square root $\qquad$ $x = \pm\sqrt{9} = \pm 3.$

Substitute $x = \pm 3$ in the equation and verify that the answers are correct.

Example 3.14.2 (from kinematics)

When an object is dropped from a certain height, its vertical distance of fall, y, during time interval t is given by the equation $y = -(9.8 \text{ m/s}^2)\, t^2$, assuming that the upward direction is positive and that t is measured in seconds. If a ball is dropped from a height of 100.0 m, when will it reach the ground?

List the data with the proper sign. Since the ball falls downward and the upward direction is assumed to be positive, in this case

$$y = -100.0 \text{ m}.$$

Use the equation $y = -(9.8 \text{ m/s}^2)\, t^2$ to solve the problem.

$$y = -(9.8 \text{ m/s}^2)\, t^2.$$

Substitute the value of y $\qquad$ $-100.0 \text{ m} = -(9.8 \text{ m/s}^2)\, t^2.$

Divide by -9.8 $\qquad$ $t^2 = \dfrac{-100.0 \text{ m}}{-9.8 \text{ m/s}^2}$

$$= 10.2 \text{ s}^2.$$

Take square root $\qquad$ $t = \pm\sqrt{10.2} \text{ s} = \pm 3.2 \text{ s}.$

positive

100 m

y

ground

The time of fall must be positive, so the only acceptable answer is $t = 3.2$ s.

General Solution of a Quadratic Equation

A *quadratic equation* of the form $ax^2 + bx + c = 0$ has the general solution

$$x = \frac{-b \pm \sqrt{b^2 - 4ac}}{2a}.$$

The two roots x_1 and x_2, are given by

$$x_1 = \frac{-b + \sqrt{b^2 - 4ac}}{2a}, \qquad x_2 = \frac{-b - \sqrt{b^2 - 4ac}}{2a}.$$

If $b^2 = 4ac,$ the roots are equal and real.

If $b^2 > 4ac,$ the roots are unequal and real.

If $b^2 < 4ac,$ the roots are complex (containing real and imaginary parts).

Although a quadratic equation may have two real unequal roots, often only one answer is physically acceptable.

Example 3.14.3

Solve the equation $2x^2 + 6x - 18 = 0$.

Compare this equation with $ax^2 + bx + c = 0$. In this case, $a = 2$, $b = 6$ and $c = -18$. Therefore,

$$x = \frac{-b \pm \sqrt{b^2 - 4ac}}{2a}$$

Substitute the values

$$= \frac{-6 \pm \sqrt{6^2 - 4 \cdot 2(-18)}}{2 \cdot 2} = \frac{-6 \pm \sqrt{36 + 144}}{4}$$

$$= \frac{-6 \pm \sqrt{180}}{4} = \frac{-6 \pm 6\sqrt{5}}{4} = \frac{-3 \pm 3\sqrt{5}}{2}.$$

Thus, the roots are

$$x_1 = \frac{-3 + 3\sqrt{5}}{2}, \qquad x_2 = \frac{-3 - 3\sqrt{5}}{2}.$$

Example 3.14.4 (from kinematics)

The following equation describes how the position y of an object undergoing a uniform acceleration a depends on time t: $(y - y_0) = v_0t + \frac{1}{2}at^2$. Here, v_0 is the initial speed of the object and y_0 is its initial position.

A ball is thrown downward from the top of a 20.0 m high building with an initial speed of 2.0 m/s. The acceleration of such a freely falling object near the earth is known to be 9.8 m/s². Determine the time when the ball will reach the ground.

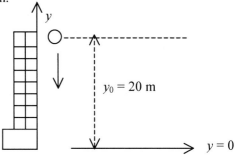

Consider the y-axis to be pointed upward.

Then, $y_0 = +20.0$ m,

$v_0 = -2.0$ m/s,

$a = -9.8$ m/s², and

$y = 0$ (since the ball reaches the ground).

Use the equation $(y - y_0) = v_0t + \dfrac{1}{2}at^2$ to solve the problem. Rearranging the equation,

we get

$$\frac{1}{2}at^2 + v_0t - (y - y_0) = 0$$

Substitute the values of y_0, v_0, y, and a

$$\frac{1}{2} \times (-9.8)t^2 + (-2.0)t - (0 - 20.0) = 0$$

or

$$4.9\,t^2 + 2t - 20 = 0.$$

This is a quadratic equation in t. Compare it with $ax^2 + bx + c = 0$. In this case, $a = 4.9$, $b = 2$, and $c = -20$.

The solutions are

$$t = \frac{-2 \pm \sqrt{2^2 - 4 \times 4.9\,(-20)}}{2 \times 4.9}$$

$$= \frac{-2 \pm \sqrt{4 + 392}}{9.8} = \frac{-2 \pm 19.9}{9.8}$$

$$= 1.8 \text{ s or } -2.2 \text{ s.}$$

The answer $t = -2.2$ s is not an acceptable solution because the ball must have fallen after it was dropped (i.e., the time cannot be negative). Thus, the ball reaches the ground in 1.8 seconds.

Exercise

29. Solve the following quadratic equations

    a) $-200 = -16t^2$ b) $3t^2 - 2t - 3 = 0$ c) $y^2 + y + 1 = 0$

30. An object is dropped from a height of 40 m. When will the object reach the ground? Assume that its vertical position y is related to its time of fall t by $y = v_0 t + \frac{1}{2} at^2$. The acceleration a is directed downward and is 9.8 m/s^2 in magnitude.

3.15   Inequalities

    An *inequality* is a mathematical statement of the fact that two quantities are not, or may not be, equal. For example, $x + 2 \geq 7$ is an inequality. Any value of x that satisfies the inequality is a solution of the inequality. Usually, an inequality has many solutions.

Assume that x, y, and a are real numbers, and use the following addition, subtraction, multiplication, and division properties of an inequality:

Addition and subtraction property	If $x < y$, then $x \pm a < y \pm a$
Example:	$2 < 4$, so if we add 3 on both sides, we get $2 + 3 < 4 + 3$, or $5 < 8$, as expected.
Multiplication property	If $x < y$ and $a > 0$, then $xa < ya$. If $x < y$ and $a < 0$, then $xa > ya$.
Example:	$2 < 4$, so if we multiply by 3, we get, $2 \cdot 3 < 4 \cdot 3$, or $6 < 12$, as expected.
Division Property	If $x < y$ and $a > 0$, then $\dfrac{x}{a} < \dfrac{y}{a}$. If $x < y$ and $a < 0$, then $\dfrac{x}{a} > \dfrac{y}{a}$.

Example: 2 < 4, divide by 2, we get 2/4 < 4/2, that is, ½
                                           < 2, as expected.
                                           Also, 2 < 4, divide by –2, we get 2/–2 > 4/–2,
                                           that is, –1 > –2, as expected.

Similar statements hold for >, ≥, and ≤.

Note: If both sides of an inequality are multiplied or divided by a *negative* number, the direction of the inequality is *reversed*.

Example 3.15.1

Solve $3 - 2x \leq 11$.

$$3 - 2x \leq 11$$

Add –3 on both sides $- 2x \leq 11 - 3$

or $-2x \leq 8.$

Divide by –2 $x \geq -\dfrac{8}{2}$ (note the change of the

                                                         direction of the inequality)

or $x \geq -4.$

Check answer by placing a number, such as $x = -3$ in the original inequality to verify the solution.

Example 3.15.2 (from thermal physics)

Room temperature is expressed in the Fahrenheit scale (F) in the United States and in the Celsius scale (C) in Canada. The temperature of a room in the United States is maintained between 68° F and 71° F. What would be the corresponding range in the Celsius scale? The temperatures in the two scales are related by the equation $F = \dfrac{9}{5}C + 32$. We have

$$68 \leq F \leq 71.$$

Substitute $F = \dfrac{9}{5}C + 32$ $68 \leq \dfrac{9}{5}C + 32 \leq 71.$

Add –32 $68 - 32 \leq \dfrac{9}{5} C \leq 71 - 32.$

Multiply by 5/9 $36 \cdot \dfrac{5}{9} \leq C \leq 39 \cdot \dfrac{5}{9},$

                                             $20 \leq C \leq 21.7$

Thus, the temperature of the room must be maintained between $20°\,C$ and $21.7°\,C$.

Exercises

31. Solve

    a) $2x + 7 > x - 2$ b) $x - 4(x - 1) > 11$ c) $-4 < 3(x - 1) \leq 10$

32. In the first two physics tests, a student has test scores of 88% and 87%. What must he score on the third test to earn 90% or more cumulatively?

Logarithms and the Exponential Function

3.16 Logarithms

If $y = b^x$, then x is called the *logarithm* of y relative to b. This is written as $\log_b y = x$. Thus, the *logarithm* of a positive number, y, relative to a positive base, b (different from 1), is defined as the exponent or the power to which the base b is raised to produce the number y.

The following rules apply to logarithms:

$$\log_b (mn) = \log_b m + \log_b n \qquad \text{(rule for addition)}$$

$$\log_b \left(\frac{m}{n} \right) = \log_b m - \log_b n \qquad \text{(rule for subtraction)}$$

$$\log_b (m^n) = n \log_b m \qquad \text{(rule for powers)}$$

$$(\log_a m)(\log_m n) = \log_a n \qquad \text{(rule for switching base)}$$

The preceding rules can easily be verified by choosing any base b (different from 1) and any positive numbers m and n. Show that the left-hand side and right-hand sides are equal.

For example,   let us choose $b = 10$, $m = 3$ and $n = 2$. Then,

$$\log_b (mn) = \log_{10}(3 \cdot 2) = \log_{10} 6 = 0.77815 \text{ (result from a calculator)}.$$

Also, $\log_b m + \log_b n = \log_{10} 3 + \log_{10} 2 = 0.47712 + 0.30103 = 0.77815$.

Therefore, $\log_{10}(3 \cdot 2) = \log_{10} 3 + \log_{10} 2$.

This verifies the addition rule: $\log_b (mn) = \log_b m + \log_b n$.

Similarly, other rules can easily be verified.

Example 3.16.1

Write the expression $\log_b \dfrac{x\,y^4}{z^2}$ in terms of logarithms of x, y, and z.

Use the rule for subtraction $\log_b \dfrac{x\,y^4}{z^2}$ $= \log_b (x\,y^4) - \log_b z^2$

Use the rule for addition $= (\log_b x + \log_b y^4) - \log_b z^2$

Use the rule for powers $= \log_b x + 4 \log_b y - 2 \log_b z.$

Example 3.16.2

Simplify the expression $\log_b\left(\dfrac{x}{y} + x\right) - \log_b\left(\dfrac{z}{y} + z\right)$. We have

$$\log_b\left(\frac{x}{y} + x\right) - \log_b\left(\frac{z}{y} + z\right)$$

Use the rule for subtraction $= \log_b \dfrac{\left(\dfrac{x}{y} + x\right)}{\left(\dfrac{z}{y} + z\right)}$

Multiply the numerator and denominator by y $= \log_b \dfrac{(x + xy)}{(z + zy)}$

Factor out x and z $= \log_b \dfrac{x(1 + y)}{z(1 + y)} = \log_b \dfrac{x}{z}.$

Exercises

33. Use the properties of logarithms and write each expression in terms of logarithms of x, y, and z.

    a) $\log_b(x^3 y^2 z^{10})$ b) $\log_b \sqrt[3]{\dfrac{x^2 y^3}{z^2}}$ c) $\log_b \sqrt{x^2 y^4}$

34. Simplify each expression.

a) $4 \log_b x + \dfrac{1}{2} \log_b y - 2 \log_b(x + y)$

b) $\log_b(xy + x^2) + \log_b(xz + yz) - \log_b z$

Common Logarithms

For many logarithmic computations, it is common to use $b = 10$ as the base. In this case, the logarithm is called a *common logarithm*. Usually, $\log y$ means $\log_{10} y$. Since logarithm with respect to base-10 appears very often, it is useful to remember that

$\log 1 = 0$	because	$10^0 = 1,$
$\log 10 = 1$	because	$10^1 = 10,$
$\log 100 = 2$	because	$10^2 = 100,$
$\log 1000 = 3$	because	$10^3 = 1000,$
$\log (0.1) = -1$	because	$10^{-1} = 0.1,$
$\log (0.01) = -2$	because	$10^{-2} = 0.01,$
$\log (0.001) = -3$	because	$10^{-32} = 0.001.$

Before calculators were invented, extensive tables provided logarithms of numbers. It is now very easy to calculate the logarithm of a number using a calculator. You should read the manual of your calculator to calculate logarithms and determine its full abilities. Usually, you type the number, press the LOG key, and then press the key that has sign '=', the equals sign. For example, to find log 32.58, enter 32.58 and press LOG. To use a graphing calculator, enter LOG 32.58 and then ENTER. To determine a number whose logarithm is known, usually you enter the known number, press the 10^x key, and then press the "=" key. The plot of $y = \log_{10} x$ is shown in Figure 3.6.

The base-10 logarithm is used often in physics, such as to define the sound intensity level in acoustics and the voltage gain in electronics. The logarithmic unit of sound intensity level or voltage gain is called the decibel (dB). In geophysics, the base-10 logarithm is used to measure the intensity of earthquakes on the Richter scale. The base-10 logarithm is also used in chemistry to determine the hydrogen ion concentration of a solution. The next several examples and exercises will illustrate the use of the base-10 logarithm in these cases.

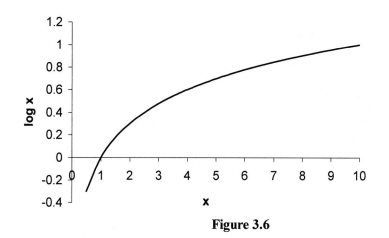

Figure 3.6

Example 3.16.3 (from acoustics)

The base-10 logarithm is used in physics to determine the sound intensity level in decibels (dB), which is a measure of the loudness of sound. The sound intensity level β is defined as $\beta = 10 \log \dfrac{I}{I_0}$, where I is the intensity of sound in watts/square meter (W/m^2) and $I_0 = 10^{-12}$ W/m^2, called the threshold intensity (minimum sound intensity) of hearing. Determine the sound intensity level at a point where the sound intensity is $I = 10^{-4}$ W/m^2.

For this problem, $I = 10^{-4}$ W/m^2

Also, $I_0 = 10^{-12}$ W/m^2

$$\beta = ?$$

Following is a sketch of the problem:

$\vartriangleleft$ - $\bullet$ $I = 10^{-4}$ W/m^2

Use the expression for sound level, $\beta = 10 \log \dfrac{I}{I_0}$, to solve the problem. Unit of β is dB.

Therefore, $$\beta = 10 \log \frac{I}{I_0}$$

Substitute the values of I and I_0
$$= 10 \log \frac{10^{-4} \, \text{W}/\text{m}^2}{10^{-12} \, \text{W}/\text{m}^2} = 10 \log (10^8) \text{ dB}$$

Use the rule for powers
$$= 10 \times 8 \log 10 \text{ dB } [\text{since } \log (m^n) = n \log m]$$

$$= 80 \log 10 \text{ dB}$$

$$= 80 \times 1 \text{ dB} \qquad [\text{since } \log 10 = 1]$$

$$= 80 \text{ dB}$$

Example 3.16.4 (from acoustics)
The sound intensity level at a point is 91.0 dB. Determine the sound intensity.

For this problem, $$\beta = 91.0 \text{ dB}$$

$$I_0 = 10^{-12} \text{ W/m}^2$$

$$I = ?$$

Therefore, $$\beta = 10 \log \frac{I}{I_0}$$

Substitute the values of β and I_0
$$91.0 \text{ dB} = 10 \log \frac{I}{10^{-12} \, \text{W}/\text{m}^2}$$

Divide by 10
$$\log \frac{I}{10^{-12} \, \text{W}/\text{m}^2} = \frac{91}{10} = 9.1$$

$$\frac{I}{10^{-12} \, \text{W}/\text{m}^2} = \text{INV } \log (9.1) \text{ [which}$$

$$\text{is } 10^{9.1}]$$

Take the inverse logarithm
$$= 1.26 \times 10^9 \text{ W/m}^2$$

Multiply by 10^{-12}
$$I = 10^{-12} \times (1.26 \times 10^9) \text{ W/m}^2$$

$$= 11.3 \times 10^{-3} \text{ W/m}^2.$$

Example 3.16.5 (from electronics)

The base-10 logarithm is also used in electronics to determine the voltage gain of an electronic device, such as an electronic amplifier. The voltage gain is expressed in decibel. If V_0 is the output voltage and V_i is the input voltage of an amplifier, its decibel voltage gain is given by $20 \log \dfrac{V_0}{V_i}$. Determine the decibel voltage gain of an amplifier that provides an output voltage of 56 V when the input voltage is 0.11 V.

For this problem, $V_0 = 56$ V,

$V_1 = 0.11$ V.

Use the expression voltage gain $= 20 \log \dfrac{V_0}{V_i}$ to determine the voltage gain in decibel (dB).

$$\text{Voltage gain } = 20 \log \frac{V_0}{V_i} = 20 \log \frac{56 \text{ V}}{0.11 \text{V}} = 20 \log (509.1)$$
$$= 20 \times 2.7 = 54 \text{ dB}.$$

Example 3.16.6 (from chemistry)

To express the acidity of a solution, the common (base-10) logarithm is used. The concentration is expressed by pH, which is defined as $pH = -\log [H^+]$, where H^+ is the hydrogen ion concentration of the solution in gram-ions per liter. Determine the hydrogen ion concentration of a solution if its pH is 8.2.

We have   pH = 8.2.

Use the equation $pH = -\log [H^+]$ to calculate the hydrogen ion concentration [H+].

$$pH = -\log [H^+].$$

Substitute the value of pH $8.2 = -\log [H^+]$

or $\log [H^+] = -8.2$

$[H^+] = $ Inv log (-8.2).   (This is $10^{-8.2}$.)

Take the inverse logarithm $[H^+] = 6.3 \times 10^{-9}$ gram-ions per liter.

(Use a calculator.)

Exercises

35. Determine the sound intensity at a point where the sound intensity level is 50 dB.

36. The threshold of pain (the maximum sound that you can hear without causing any problem) is 120 dB. Determine the corresponding sound intensity, whose unit is watt per square meter (W/m^2).

37. Determine the voltage gain of an electronic amplifier whose input and output voltages are 2.0 mV and 105.0 mV.

38. The pH of pure water is 7. Determine the hydrogen ion concentration of pure water in gram-ions per liter.

39. Find the pH of a solution if its hydrogen ion concentration is 4.6×10^{-7} gram-ions per liter.

40. In seismology the intensity of an earthquake on the Richter scale R is expressed as $R = \log \dfrac{A}{T}$, where A is the amplitude (measured in micrometers) and T is the period (the time of vibration of the earth's surface, measured in seconds.) Determine the measure of an earthquake on the Richter scale with amplitude $A = 10,000$ micrometers and period $T = 0.15$ s.

Natural logarithms

In many applications another base, called the natural base $e = 2.718...$, is used. Mathematically, the number e is the value of the limit, $\lim_{n \to \infty} (1 + 1/n)^n$, which is equivalent to $\lim_{z \to 0} (1 + z)^{1/z}$, where $z = 1/n$. The preceding limits mean that if you calculate the value of $(1 + 1/n)^n$ for different values of n (or the value of $(1 + z)^{1/z}$ for values of z), you will find that as n becomes larger and larger (or z becomes closer and closer to zero), the value of $(1 + 1/n)^n$ [or the value of $(1 + z)^{1/z}$] will become 2.718... For example, when

$$z = 1, \qquad (1 + z)^{1/z} = 2.0,$$
$$z = 0.5 \qquad (1 + z)^{1/z} = 2.25,$$
$$z = 0.25 \qquad (1 + z)^{1/z} = 2.4414,$$
$$z = 0.1 \qquad (1 + z)^{1/z} = 2.5937,$$
$$z = 0.01 \qquad (1 + z)^{1/z} = 2.7048,$$

$$z = 0.001 \qquad (1 + z)^{1/z} = 2.7169,$$
$$z = 0.0001 \qquad (1 + z)^{1/z} = 2.7181.$$

Clearly, the expression $(1 + z)^{1/z}$ approaches the value 2.718... as z becomes closer and closer to zero.

Example 3.16.7

If A_0 dollars are put into an account that pays an annual interest rate of r, compounded N times in one year, it can be shown that the amount of money after t years becomes $A(t) = A_0 (1 + r/N)^{Nt}$. If the interest is compounded continuously, show that $A(t) = A_0 e^{rt}$. Determine the balance in the account after five years at the 4% rate of interest if the initial deposit A_0 is $100.

Since the interest is compounded continously, it follows that $N = \infty$, and the balance after t years is

$$
\begin{aligned}
A(t) \quad &= \lim_{N \to \infty} A_0 (1 + r/N)^{Nt} \\[2mm]
&= A_0 \lim_{n \to \infty} [(1 + r/N)^{N\,r}]^{rt}
\end{aligned}
$$

Substitute $n = N/r$
$$= A_0 [\lim_{n \to \infty} (1 + 1/n)^{n}]^{rt} \quad [\text{ as } N \to \infty \, , n \to \infty].$$

Substitute e for the limit
$$= A_0 \, e^{rt}.$$

For the next part of the problem substitute $A_0 = \$100$, $r = 4\% = 0.04$, and $t = 5$ years

Therefore,
$$A(t) \quad = A_0 \, e^{rt} = \$100 \, e^{0.04 \times 5} = \$122.14.$$

Likewise, the common logarithm is defined when $y = 10^x$, and the logarithm with respect to the base e, known as *natural logarithm* and denoted by ln x, is defined when $y = e^x$. The exponential function e^x and natural logarithm occur in several natural phenomena where the rate of change of some quantity depends on the quantity. A plot of $y = \ln (x)$ is shown in Figure 3.7.

Note that $e^{\ln x} = x$. This can be easily verified by taking the natural logarithm of $e^{\ln x}$.

$$\ln (e^{\ln x})$$

Use the power rule $= \ln x \ln e$

By definition, $\ln e = 1$ $= \ln x \cdot 1$

                                       $= \ln x$

Therefore, $e^{\ln x} = x$. Because of their inverse relationship, e^x and $\ln x$ are called *inverse functions*.

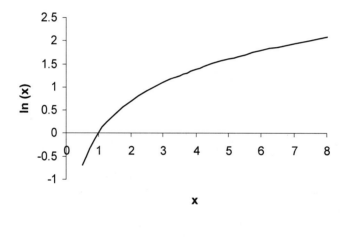

Figure 3.7

3.17   Exponential Functions

Exponential functions play a special role in mathematics and physics. The inverse of the natural logarithm, $\ln x$ is the exponential function, $y = e^x$, also written as *exp* (*x*). Similarly, the negative exponent exponential function is $y = e^{-x}$, also written as *exp* (*–x*). The plots of these exponential functions are shown in Figure 3.8 and Figure 3.9, respectively, for $x = -1$ to 2.

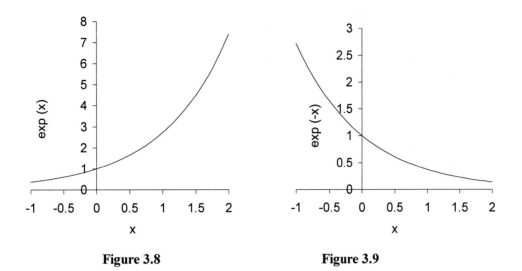

Figure 3.8   **Figure 3.9**

When a quantity increases like the positive exponent exponential function, it is called *exponential growth*.   Similarly, when a quantity decreases like the negative exponent exponential function, it is called *exponential decay*.

From the plot of the function y = exp (x) (that is, e^x), it is easily seen that the function is a curve which has small values when its slope is small and large values when the slope is large.   Thus, the rate of change of the curve $y = e^x$ is e^x.   Similarly, the rate of decrease of the curve $y = e^{-x}$ is e^{-x}.   This property is special in physics, since it occurs naturally, such as in the radioactive decay of an unstable nucleus of an atom. For radioactive decay, the rate of decay of the number of nucleus, N, with time t is proportional to N.   This can be expressed as $\Delta N/\Delta t \propto N$.   Thus, the number decays exponentially. Note that when there is sufficient food and space, populations of living organisms also tend to increase exponentially.

Note: The exponential function e^x and e^{-x} and natural logarithm occur in several natural phenomena in which the *rate of change (increase or decrease) of some quantity depends on the quantity.*

---

74   Chapter 3

Example 3.17.1 (from nuclear physics)

Radon is radioactive and it decays. The number N of radon atoms remaining after time t (in days) is related to the original number N_0 of radon atoms by the expression $N = N_0\, e^{-(0.181/\text{day})t}$. If $N_0 = 2 \times 10^7$, find the time after which 2×10^5 radon atoms will remain.

For this problem, $N = 2 \times 10^5$
$$N_0 = 2 \times 10^7.$$

Use the equation $N = N_0\, e^{-(0.181/\text{day})}$ to determine the time t. Note that the power of e must be dimensionless. Thus, the unit of the time t is "day." We have

$$N = N_0\, e^{-(0.181/\text{day})t}.$$

Substitute the values of N and N_0 $2 \times 10^5 = 2 \times 10^7\, e^{-(0.181/\text{day})t}.$

Divide by 2×10^7 $\dfrac{2 \times 10^5}{2 \times 10^7} = e^{-(0.181/\text{day})t}.$

or $10^{-2} = e^{-(0.181/\text{day})t}.$

Change the equation into logarithmic form $\ln(10^{-2}) = -(0.181/\text{day})t.$

Use the rule for powers, $\ln(m^n) = n \ln m$ $-2 \ln 10 = -(0.181/\text{day})t.$

Divide by -0.181 $t = \dfrac{2 \ln 10}{0.181/\text{day}}$

$$= \dfrac{2 \times 2.303}{0.181/\text{day}} = 25.4 \text{ days.}$$

Exercises

41. The number of atoms, N, that remain of an unstable radioactive atom is given by the expression $N = N_0\, e^{-\lambda t}$. If the initial number of atoms is $N_0 = 2.4 \times 10^{15}$ and after 20 days the number of remaining atoms is $N = 1.2 \times 10^{14}$, determine the constant λ. The constant λ, called the decay constant, has unit day^{-1}.

42. The growth of bacteria in a laboratory culture increases according to the expression $N = N_0\, e^{(0.37\,/\text{s})\, t}$. If the initial population is 500, determine the time when the population will be 5000.

The Binomial Theorem and Series Expansion

3.18   The Binomial Theorem

Consider the following *binomial expansions*, which you can verify by direct multiplication,

$$(a + b)^2 = (a + b)(a + b) = a^2 + ab + ba + b^2 = a^2 + 2ab + b^2$$

$$(a + b)^0 = 1$$

$$(a + b)^1 = a + b$$

$$(a + b)^2 = a^2 + 2ab + b^2$$

$$(a + b)^3 = a^3 + 3a^2b + 3ab^2 + b^3$$

$$(a + b)^4 = a^4 + 4a^3b + 6a^2b^2 + 4ab^3 + b^4$$

In general, according to the *binomial theorem*, for any positive integer n,

$$(a + b)^n = a^n + \frac{n!}{1!(n-1)!}a^{n-1}b + \frac{n!}{2!(n-2)!}a^{n-2}b^2 + \frac{n!}{3!(n-3)!}a^{n-3}b^3 + \ldots$$

$$+ \frac{n!}{p!(n-p)!}a^{n-p}b^p + \ldots + b^n$$

$$= \sum_{p=0}^{n} \frac{n!}{(n-p)!\,p!}a^{n-p}b^p$$

In the preceding expressions, the symbol ! is called factorial. $n!$ is defined as
$n! = n(n-1)(n-2)\ldots\ldots 3 \cdot 2 \cdot 1$.

For example, $6! = 6 \cdot 5 \cdot 4 \cdot 3 \cdot 2 \cdot 1 = 720$.

Also, $n! = n(n-1)(n-2)\ldots\ldots 3 \cdot 2 \cdot 1 = n[(n-1)(n-2)\ldots\ldots 3 \cdot 2 \cdot 1] = n(n-1)!$.

Example 3.18.1

Expand $(3x + 2y)^4$, using the binomial theorem.

$$(3x + 2y)^4 \quad = (3x)^4 + \frac{4!}{1!(4-1)!}\,(3x)^{4-1}(2y) + \frac{4!}{2!(4-2)!}\,(3x)^{4-2}(2y)^2$$

$$+ \frac{4!}{3!(4-3)!}\,(3x)^{4-3}(2y)^3 + (2y)^4$$

$$= (3x)^4 + 4\,(3x)^3(2y) + 6\,(3x)^2(2y)^2 + 4\,(3x)\,(2y)^3 + (2y)^4$$

$$= 81x^4 + 216x^3y + 216x^2y^2 + 96xy^3 + 16y^4$$

Exercise

43. Use the binomial theorem to expand these expressions:

i) $(a + b)^5$

ii) $\left(\dfrac{a}{3} - \dfrac{b}{2}\right)^4$

3.19   Power Series Expansions and Approximations

The right-hand side of the expressions that will follow usually contains an infinite number of terms. A polynomial with an infinite number of terms is called a *power series*; hence, the following expansions are called *power series expansions*.

$$(a + b)^r = a^r + \frac{r}{1!}\,a^{r-1}b + \frac{r(r-1)}{2!}\,a^{r-2}b^2 + \dots \qquad e^x = 1 + \frac{x}{1!} + \frac{x^2}{2!} + \frac{x^3}{3!} + \dots$$

$$(1 + x)^r = 1 + rx + \frac{r(r-1)}{2!}\,x^2 + \dots$$

$$\ln(1 \pm x) = \pm x - \frac{1}{2}x^2 \pm x^3 + \dots$$

(In the preceding expressions that involve r, r is a negative integer or a fraction.)

It is often desirable *to approximate* the solution of a problem, either because the exact solution may be difficult to obtain or may not be necessary, or because we may need the value for a limiting case, say, when the variable x is very small. The preceding power series expansions are very helpful for such approximations. For example, when x is very small ($x \ll 1$), the x^2 and other higher order terms are negligibly small and

$$(1 + x)^r \approx 1 + rx$$

$$e^x \approx 1 + x,$$

$$\ln(1 \pm x) \approx \pm x.$$

Example 3.19.1 (from special relativity)

The ability do work is called energy. Kinetic energy is the energy of motion. For a spaceship moving at a speed v, its kinetic energy is given by $K = mc^2 \left(\dfrac{1}{\sqrt{1 - v^2/c^2}} - 1 \right)$ where c is the speed of light. Show that, for small velocity ($v \ll c$), kinetic energy of the spaceship becomes $\dfrac{1}{2} mv^2$.

spaceship

We have

$$K = mc^2 \left(\frac{1}{\sqrt{1 - v^2/c^2}} - 1 \right) = mc^2 \left[\left(1 - v^2/c^2\right)^{-1/2} - 1 \right]$$

Now, $(1 + x)^n \approx 1 + nx$; therefore, $\left(1 - v^2/c^2\right)^{-1/2} \approx 1 - \left(-\dfrac{1}{2} \right) \dfrac{v^2}{c^2}$ [$x = -v^2/c^2$ and $n = -\frac{1}{2}$].

Thus, the preceding expression for kinetic energy can be written as

$$K = mc^2 \left(\frac{1}{\sqrt{1 - v^2/c^2}} - 1 \right).$$

$$\approx mc^2 \left(\left(1 - \left(-\frac{1}{2} \right) \frac{v^2}{c^2} \right) - 1 \right)$$

$$= mc^2 \frac{1}{2} \frac{v^2}{c^2} = \frac{1}{2} mv^2.$$

Exercises

44. According to the relativistic time dilation formula, if a car travels at speed v with respect to the earth, the time t_0 recorded by the driver according to his own watch is related to the time t recorded by an observer on the earth by

$$t = \frac{t_0}{\sqrt{1 - v^2/c^2}}, \quad \text{where } c = 3 \times 10^8 \text{ m/s is the speed of light. If the speed of}$$

the car is 110 km/hr, and the driver travels for 10 seconds according to his watch, show that the difference between t and t_0 is about 5×10^{-14} seconds.

[*Hint*: First, show that the relation can be approximated as $t = t_0 (1 + \frac{1}{2} v^2/c^2)$. Also, you must convert the speed of the car to meters per second.]

45. According to the general theory of relativity, the speed of light is reduced in the space surrounded by an astronomical body (such as a star) of mass M and radius r by the factor $\sqrt{1 - 2GM/rc^2}\big/\sqrt{1 + 2GM/rc^2}$ in comparison with its speed, c, in free space. Here G is a constant, called Gravitational constant. Show that the preceding factor can be approximated as $1 - 2GM/rc^2$ when $1 - 2GM/rc^2$ is much less than 1.

46. The value of acceleration due to gravity g' at a height h from the earth's surface is given by $g' = GM/(r + h)^2$ where M and r are the mass and the radius of the earth, respectively, and G is a constant (called universal gravitational constant). If g is the acceleration due to gravity on the surface of the earth ($h = 0$ on the surface), show that when $h \ll r$, $(g' - g) \approx -2gh/r$. [*Hint*: $(r + h)^2 = r^2 (1 + h/r)^2$.]

Answers to Exercises in Chapter 3

2. 2.3×10^7 light years

3. 210 N

4. 2.43×10^5 Pascals

5. 1.8×10^4 N/m^2

6. 2.7×10^4 m/s

7. 1.8 lux

8. 0.15 times the value at the surface

9. $x^{-4}y^8z^{-7}$

10. 1

11. $x^7y^{-11}z^3$

12. $64\,a^{-9}b^{-6}c^{-6}$

13. a) $(t-3)(t-4)$

b) $2(x+4)(x-6)$

c) $6a(b+4)^2$

d) $(2a+3)(a+1)$

14. $x = -2$

15. $x = 5$

16. $x = 5$

17. $G = Fr^2/Mm$

18. $P = RT/(V-b) - a/V^2$

19. $\alpha = (L - L_0)/(L_0\Delta T)$

20. $f = d_0d_i/(d_0 + d_i)$

21. 0.3927 ohm

22. 26.3 C

23. $x = 2, y = 1$

24. $x = 2, y = 3$

25. $x = 1, y = 1$

26. $x = 4, y = 2$

27. $x = 6, y = 4$

28. $x = 3, y = 4$

29. a) 3.5, –3.5

b) $t = \dfrac{1 + \sqrt{10}}{3}, \dfrac{1 - \sqrt{10}}{3}$

c) $y = \dfrac{-1 + \sqrt{-3}}{2}, \dfrac{-1 - \sqrt{-3}}{2}$

30. 2.9 s

31. a) $x > -9$

b) $x > -7/3$

c) $-1/3 < x \le 13/3$

32. 95 or more

33. a) $3\log_b x + 2\log_b y + 10\log_b z$

b) $\dfrac{2}{3}\log_b x + \log_b y - \dfrac{2}{3}\log_b z$

c) $\log_b x + 2\log_b y$

34 a) $\log_b\left[x^4 y^{1/2} \Big/ \left(x + y\right)^2\right]$

b) $\log_b[x(x+y)^2]$

35. 10^{-7} W/m^2

36. 1 W/m^2

37. 40.3 dB

38. 1.0×10^{-7} gram-ions per liter

39. 6.3

40. 4.8 on the Richter scale

41. $\lambda = 0.15$ day^{-1}

42. 6.2 s

43. i) $= a^5 + 5a^4b + 10a^3b^2 + 10a^2b^3 + 5ab^4 + b^5$

ii) $a^4/81 - 2a^3b/27 + a^2b^2/6 - ab^3/6 + b^4/16$

Chapter 4
Geometry: Dealing with Shapes and Plots

The word *geometry* is taken from two Greek words meaning the measurement of Earth. Many of the early discoveries in geometry were motivated by the need for distance and area measurements of the earth. In general, geometry deals with *the property and measurement of regular shapes* including lines, angles, curves, planes, and surfaces. The rules of geometry were first systematically studied by the Greek mathematician Euclid (300 B.C.). In physics, you will often deal with regular shapes and curves. For example, the shapes of the earth and the sun can be approximated as spheres; the orbit for the motion of an electron in a normal hydrogen atom can be approximated as a circle; in absence of air resistance, the path for the motion of any projectile (such as a soccer ball) is a parabola; etc. Knowledge of geometry is extremely useful in studying such shapes and the physical situations involving them.

Angles, Triangles, and Quadrilaterals
4.1   Angles

In plane geometry, also called Euclidean geometry, *angles* are formed when two straight lines intersect on a plane. For example, consider a box (Figure 4.1). The edges of the box meet at the corners (called the *vertex* of the angle) to form angles. Figure 4.2 shows an angle θ formed by an arc s of a circle of radius r at its center. Here the two radii intersect at the center, the vertex of the angle. An angle is considered to be positive if it is measured counterclockwise from the positive x-axis, and it is negative if measured clockwise from the positive x-axis.

An angle is measured in *degrees* or in *radians*. The most common unit of measure for angle is the *degree*. If the circumference of a circle is divided equally into 360 parts, each part would make a 1-degree angle at the center of the circle. The ratio of the length of the

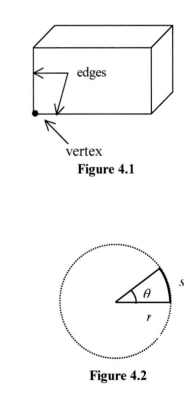

Figure 4.1

Figure 4.2

circular arc, *s*, to the radius *r* of a circle is the measure of the angle in *radians* subtended by the arc at the center of the circle. From Figure 4.2,

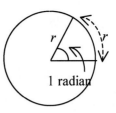

$$\theta = \frac{s}{r} \text{ in radians.}$$

Figure 4.3

From the preceding equation, when $s = r$, θ is 1 radian. Thus, a *radian* is the angle subtended at the center of a circle by an arc whose length is equal to the radius of the circle (Figure 4.3). Since the circumference of a circle is $2\pi r$, a complete circle has an angular measure of 2π radians, also called 360 degrees (°). Thus, $360° = 2\pi$ radians or $1° = \pi/180$ radian. Hence, 1 radian = $180/\pi = 57.3°$, and further,

$45° = \pi/4$ radians	$90° = \pi/2$ radians
$180° = \pi$ radians	$270° = 3\pi/2$ radians
$360° = 2\pi$ radians	

In astrophysics, when talking about stars and galaxies, we often measure angles much less than a degree and use *seconds of arc* to measure angles. A degree is divided into 60 minutes of arc, each of which is further divided into 60 seconds of arc. Thus, $1° = 3600$ seconds of arc. 1 radian = $57.3°$ degrees = 57.3 X 3600 arcseconds = 2.06×10^5 arcseconds.

Example 4.1.1

The distance between the sun and the earth is 1.5×10^{11} m. The sun makes an angle of about $0.4°$ to us on the earth. Use this information to estimate the radius of the sun.

List the data. Change the angle to radians.

Distance, $r = 1.5 \times 10^{11}$ meters

$\theta = 0.4° = 0.4 \times \pi/180$ radian

For small angles, the arc length and the chord length (the diameter of the sun, in this example) are approximately equal. Use the relation $s = r \cdot \theta$ to determine the diameter of the Sun.

$$s = r \cdot \theta$$

Substitute the values of r and θ $s = (1.5 \times 10^{11} \text{ m}) \times 0.4 \times \pi/180 \text{ radian})$

$$= 10.5 \times 10^8 \text{ m}$$

$$= 1.1 \times 10^9 \text{ m}.$$

Exercise

1. A bird can distinguish objects that subtend an angle of at least 0.02° to its eye.   What is the angle in radians? When the bird is flying at a height of 120 m, how small an object can it distinguish?

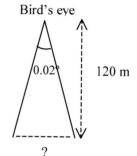

Bird's eye

0.02°

120 m

?

Classification of Angles

When two lines meet to form one straight line, it makes an angle of 180° (Figure 4.4 a). If an angle is 90°, it is called a *right angle* (Figure 4.4 b).

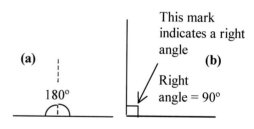

(a)

180°

This mark indicates a right angle

(b)

Right angle = 90°

Figure 4.4

An angle is called *acute* if its value is between 0° and 90° (Figure 4.5 a). An angle is called *obtuse* if its value is between 90° and 180° (Figure 4.5 b).

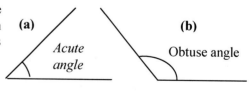

(a)

Acute angle

(b)

Obtuse angle

Figure 4.5

If two angles sum to 90° (such as angles A and B in Figure 4.6 a), they are called *complementary angles*. If two angles sum to 180° (such as angles A and B in Figure 4.6 b), they are called *supplementary angles*.

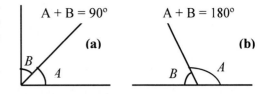

Figure 4.6

Equality of Angles

Some common situations in physics, when two angles can be equal, are shown in Figures 4.7 a–d.

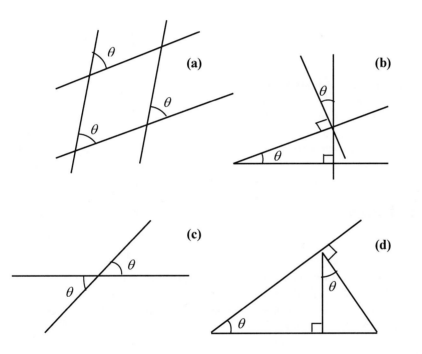

Figures 4.7 (a–d)

From Figures 4.7 (a–d), note that two angles are equal when any of the following are true:

i)   The angles are made by a line crossing parallel lines, as shown in Figure 4.7 a.

ii)   The two sides of the angles are perpendicular, as shown in Figure 4.7 b.

iii)   The angles are opposite angles at the intersection of two lines, as shown in Figure 4.7 c.

iv)   The angles have the same complement, as shown in Figure 4.7 d.

4.2   Triangles and Quadrilaterals

When an object moves in three-dimensional space, it can move left or right, front or back, and up or down. However, in many circumstances, the motion of an object in space is confined to a two-dimensional plane. For example, when you throw a ball, its motion is confined to a plane; the motion of a planet around the sun and the motion of a satellite orbiting the earth are also confined to a plane. Two-dimensional geometry is useful to describe the physics of these objects. The two-dimensional or plane figures that are common in physics, such as triangles and quadrilaterals, are described next.

Triangles

A *triangle* is a figure formed by the intersections of three straight lines on a plane. A triangle has three sides and three internal angles. The three sides and the three angles can be different. Also, it is possible that the two sides and two angles or all three sides and three angles be equal.

Equilateral Triangle: An equilateral triangle is a triangle that has equal sides and each angle equal to 60°, as shown in the Figure 4.8 a.

Isosceles Triangle: An isosceles triangle is a triangle that has two equal

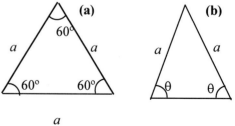

Figure 4.8

sides and two equal angles, as shown in Figure 4.8 b.

Right Triangle: A right triangle (Figure 4.9a) is a triangle that has one angle 90°. In a *right isosceles triangle* (Figure 4.9 b), the acute angles are each 45°.

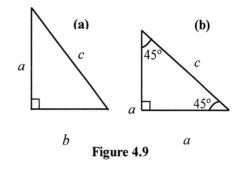

Figure 4.9

The length of the base, *b*, height, *a*, and hypotenuse, *c*, of a right triangle are related by the *Pythagorean theorem*:

$$c^2 = a^2 + b^2.$$

30–60–90 *Triangle*: If you split an equilateral triangle exactly in half, each part becomes a 30–60–90 triangle (Figure 4.10). That is, the angles of the resulting triangles are 30°, 60°, and 90°. In this case, the length of the side opposite the 30° angle is one-half the length of the hypotenuse.

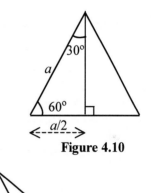

Figure 4.10

Similar Triangles: Similar triangles are triangles that have the same shapes, but that differ only in size or orientation, as shown in Figure 4.11. The internal angles of two similar triangles are equal and the ratios of the corresponding sides are equal. If the sides are *a*, *b*, and *c* and *a′*, *b′*, and *c′*, respectively, for the two similar triangles, then

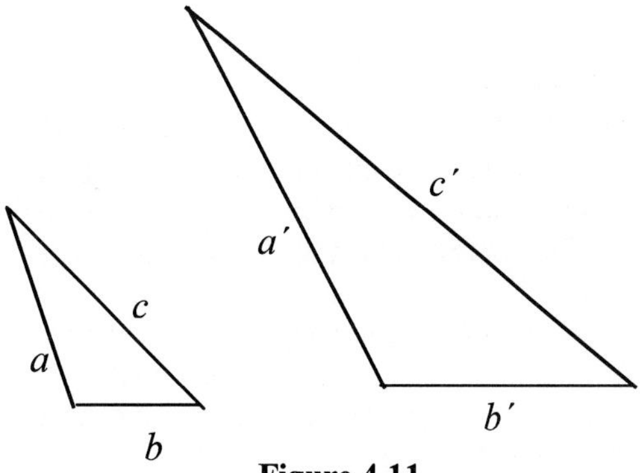

Figure 4.11

$$\frac{a}{a'} = \frac{b}{b'} = \frac{c}{c'}.$$

Congruent Triangles: Two triangles are said to be *congruent* if they have exact fit when one is placed on the top of the other. That is, congruent triangles have the same size and shape. The two triangles in Figure 4.12 are congruent.

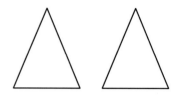

Figure 4.12

Note that in any triangle (such as that shown in Figure 4.13),

i)   the sum of the three interior angles is 180°. That is,

$$A + B + C = 180°.$$

ii)   the exterior angle, θ, is the sum of the two opposite interior angles. That is,

$$A + B = \theta.$$

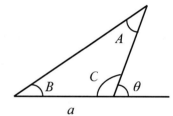

Figure 4.13

Example 4.2.1
In the figure to the right, the points D and E are the middle points of AB and CB, respectively. The angles are shown in the figure. What are the values of the angles α, β, γ, and δ? If DE is 10 cm what is the length AC?

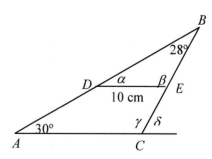

DE is parallel to AC.

$\alpha = 30°$ [since the two sides of the angles α and 30° are parallel]

$\beta = 180° - (30° + 28°)$ [summation of all angles of the triangle DBE is $180°$]

   $= 122°$.

$\gamma = 180° - (30° + 28°)$ [summation of the three angles of the triangle ABE is

$180°] = 122°$.

Note that $\beta = \gamma$, as expected, since they have two sides parallel.

$\delta = 30° + 28° = 58°$ [exterior angle is the sum of two opposite interior

angle of a triangle].

The triangles ABC and DBE are similar. Since the ratio of the sides DB and

AB is 1:2 (D is the middle point of AB), the ratio of the sides DE and AC is

1:2. That is, $AC = 2DE = 2 \times 10$ cm $= 20$ cm.

Example 4.2.2

In the diagram to the right find the angles A, B, C, and D.

$B = 50°$, because the lines forming B are
mutually perpendicular to the lines forming the
angle $50°$.
$A = 90° - B = 90° - 50° = 40°$.
$D = B$ (vertical angles) $= 40°$.
The small triangle that contains D
is a right-angled triangle.
$C = 90° - D = 90° - 40° = 50°$
(since C's vertical angle is
complementary to D).

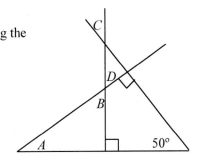

Example 4.2.3

Two tangent lines (AT_1 and AT_2) are
drawn from a point A to a circle whose
center is at O. The line OA makes an
angle of $30°$ at A with the tangent line
AT_1. Determine the angles T_2AO, AOT_1,
AOT_2, and T_1OT_2 from the figure at the
right.

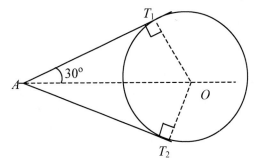

Note that a tangent line drawn to a circle at any point makes a right angle with the radius drawn from the point. That is, angles AT_1O and AT_2O are 90° each.

From symmetry, it is clearly seen that the triangles T_1AO and T_2AO are identical. Thus, the angle T_2AO is also 30°.

Angle AOT_1 = 180° − (30° + 90°) [the sum of the three internal angles of the triangle AOT_1 is 180°]

$$= 180° − 120° = 60°.$$

Angle AOT_2 = 180° − (30° + 90°) [the sum of the three internal angles of the triangle AOT_2 is 180°]

$$= 180° − 120° = 60°.$$

Angle T_1OT_2 = Angle AOT_1 + Angle AOT_2 = 60° + 60° = 120°.

Exercise

2.   For the left triangle, find the unknown angles A and B. A perpendicular has been drawn to one of the sides of the same triangle, as shown at the right. Determine the angles C and D.

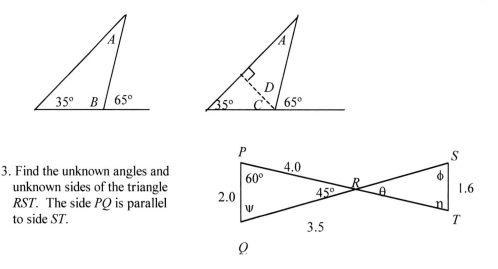

3. Find the unknown angles and unknown sides of the triangle RST. The side PQ is parallel to side ST.

Quadrilaterals

When four straight lines intersect in a plane, they form a closed figure with four sides. This figure is called a *quadrilateral*. The following quadrilaterals, where the two opposite sides are parallel to each other, as shown in Figure 4.14, are common in physics. These are called *parallelogram*, *rectangle*, and *square*, respectively. The line (dotted line for in the case of the square) that joins the two opposite corners of a quadrilateral is called its *diagonal*. Each of the figures has two diagonals. The two diagonals of a parallelogram have different lengths. The two diagonals have the same length for a rectangle or a square because of the symmetry of these two figures.

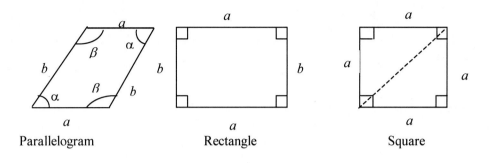

| Parallelogram | Rectangle | Square |

Figure 4.14

Note that:
- A *parallelogram* has opposite pairs of equal sides and a pair of equal angles.
- A *rectangle* is a parallelogram whose angles are all 90°.
- A *square* is a rectangle whose four sides are equal.

The sum of the four angles of any quadrilateral is 360°. Joining a diagonal, as shown in Figure 4.15, can easily verify the preceding statement. The diagonal divides the quadrilateral into two triangles, for each of which the sum of the three angles is 180°.

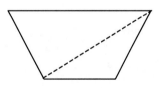

Figure 4.15

Shapes of Objects

4.3   Perimeter, Area, and Volume of Standard Shapes

A one-dimensional object (such as a very thin string) has a length, but no area or volume. A two-dimensional object (such as a piece of paper) has a perimeter and a surface area. A three-dimensional object has area for each of its surfaces and it occupies a certain amount of space, called its volume. Perimeters, areas, and volumes of common figures are often needed for solving real-world problems.

Note: A perimeter has a dimension $[L]$; that is, it has the first power in length. An area has a dimension $[L^2]$; that is, it has the second power of length; a volume has a dimension $[L^3]$, the third power of length.

It is worthwhile to remember the actual formulas for the perimeter, area, and volume of the common figures shown in Figure 4.16. These common figures include regular-shaped two-dimensional objects such as a circle and an ellipse and three-dimensional solid objects such as a rectangular parallelepiped, a cube, a sphere, and a right circular cylinder.

Two-Dimensional Figures

Figure 4.16 (a–h)

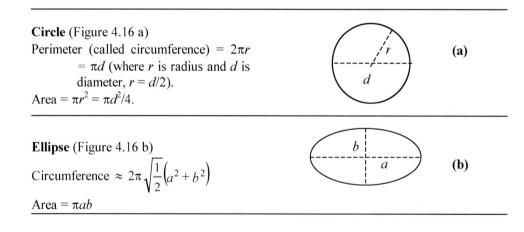

Circle (Figure 4.16 a)
Perimeter (called circumference) $= 2\pi r$
      $= \pi d$ (where r is radius and d is
      diameter, $r = d/2$).
Area $= \pi r^2 = \pi d^2/4$.

(a)

Ellipse (Figure 4.16 b)

Circumference $\approx 2\pi \sqrt{\frac{1}{2}\left(a^2 + b^2\right)}$

Area $= \pi ab$

(b)

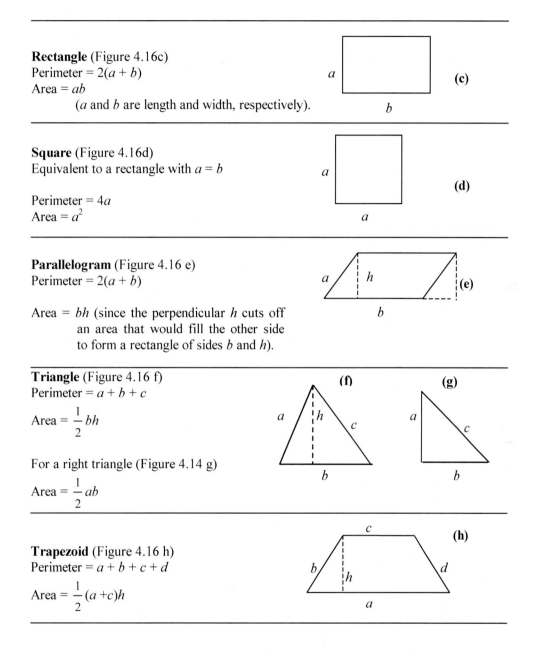

Rectangle (Figure 4.16c)
Perimeter $= 2(a + b)$
Area $= ab$
      (a and b are length and width, respectively).

(c)

Square (Figure 4.16d)
Equivalent to a rectangle with $a = b$

Perimeter $= 4a$
Area $= a^2$

(d)

Parallelogram (Figure 4.16 e)
Perimeter $= 2(a + b)$

Area $= bh$ (since the perpendicular h cuts off
      an area that would fill the other side
      to form a rectangle of sides b and h).

(e)

Triangle (Figure 4.16 f)
Perimeter $= a + b + c$

Area $= \dfrac{1}{2} bh$

For a right triangle (Figure 4.14 g)

Area $= \dfrac{1}{2} ab$

(f)   **(g)**

Trapezoid (Figure 4.16 h)
Perimeter $= a + b + c + d$
Area $= \dfrac{1}{2}(a + c)h$

(h)

Three Dimensional Figures

Rectangular parallelepiped (Figure 4.17 a) and **cube** (Figure 4.17 b)
Surface area = sum of the surface areas of all six surfaces
For a rectangular paralleopiped,
Surface area $= 2\,(ab + bc + ca)$

Volume $= abc$

For a cube, $a = b = c$

Surface area $= 6a^2$
Volume $= a^3$

(a)   **(b)**

Sphere (Figure 4.17 c)
Surface area $= 4\pi r^2$

Volume $= \dfrac{4}{3}\pi r^3 = \dfrac{1}{6}\pi d^3$ where $d = 2r$

(c)

Right circular cylinder (Figure 4.17 d)
Surface area = area of the two circular end caps plus the area of the cylindrical wrapper $= 2\pi r^2 + 2\pi r L$

Volume = the volume generated by sweeping the cross sectional area a distance L along the axis $= \pi r^2 L$.

(d)

Right circular cone (Figure 4.17 e)
Volume $= \dfrac{1}{3}\pi r^2 h$

(e)

Example 4.3.1
(a)   Calculate the circumference and area of a circle of radius 0.10 m.
(b)   Calculate the area of a triangle of height 0.20 m and base 1.40 m.
(c)   Calculate the surface area and volume of a right circular cylinder of radius 0.20 m and length 0.40 m.
(d)   Determine the volume of a cone of radius 0.20 m and height 0.80 m.

This problem is straightforward. Use the proper formula for the circumference, area, and volume to solve the problem.

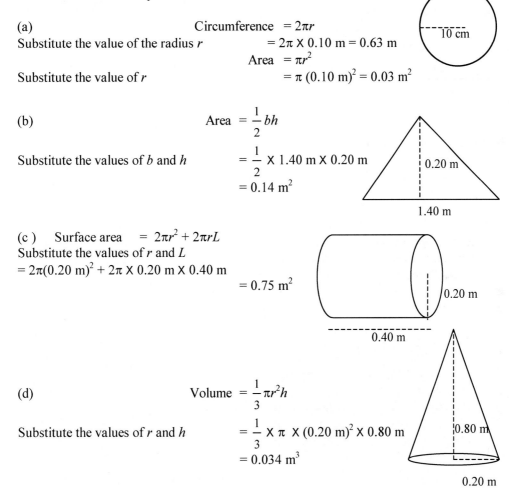

(a)   Circumference $= 2\pi r$
Substitute the value of the radius r $= 2\pi \times 0.10$ m $= 0.63$ m
                                       Area $= \pi r^2$
Substitute the value of r $= \pi (0.10$ m$)^2 = 0.03$ m^2

(b)   Area $= \dfrac{1}{2} bh$

Substitute the values of b and h $= \dfrac{1}{2} \times 1.40$ m $\times 0.20$ m

                                 $= 0.14$ m^2

(c)   Surface area $= 2\pi r^2 + 2\pi r L$
Substitute the values of r and L
$= 2\pi(0.20$ m$)^2 + 2\pi \times 0.20$ m $\times 0.40$ m
                                 $= 0.75$ m^2

(d)   Volume $= \dfrac{1}{3} \pi r^2 h$

Substitute the values of r and h $= \dfrac{1}{3} \times \pi \times (0.20$ m$)^2 \times 0.80$ m

                                 $= 0.034$ m^3

Note that, in the answers, circumference has unit m, surface area has unit m^2, and volume has unit m^3, as expected.

Example 4.3.2

The earth can be approximated to be a sphere of average radius 6.38×10^6 m. Determine the volume of the earth.

Radius $r = 6.38 \times 10^6$ m.

Use the formula for the volume of a sphere, which is $\frac{4}{3}\pi r^3$.

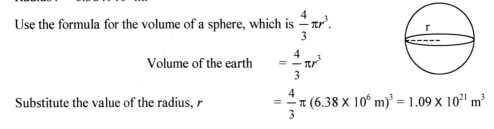

$$\text{Volume of the earth} \quad = \frac{4}{3}\pi r^3$$

Substitute the value of the radius, r

$$= \frac{4}{3}\pi\,(6.38 \times 10^6\text{ m})^3 = 1.09 \times 10^{21}\text{ m}^3$$

Exercises

4. Density. The average density D of a substance is defined as the ratio of its mass M to its volume V. Thus, $D = M/V$. If the mass of the earth is 5.97×10^{24} kg and its average radius is 6.38×10^6 m, determine the average density of the earth.

5. A circle is inscribed in a square for which sides are each 2.0 cm long and are tangents to the circle, as shown to the right. Determine the area of each corner segment.

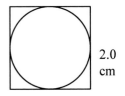

2.0 cm

2.8 cm

6. A spherical shell of metal as shown to the right has an inner diameter of 2 cm and an outer diameter of 2.8 cm. Determine the volume of copper in the shell. [*Hint*: The volume of the spherical shell is equal to the difference between the volumes of the two spheres.]

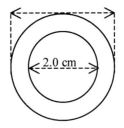

2.0 cm

The Coordinate Systems

4.4   The Cartesian Coordinate System

In geometry, coordinate systems are used to

- Find the location of an object with respect to a suitably chosen origin,

- Find the distance between two objects,

- Determine the trajectory of motion,

- And more.

In a Cartesian coordinate system (Figure 4.18) the two coordinate axes x and y are perpendicular to each other and the divisions of the x- and y-axes are equally spaced. The x- and y-axes meet at a point called the *origin*. A point P is labeled by (x, y), where magnitude of x is the distance of the point measured from the y-axis and the magnitude of y is the distance of the point measured from the x-axis. x and y can be positive or negative, depending on whether the distances are toward positive or negative axis from the origin.

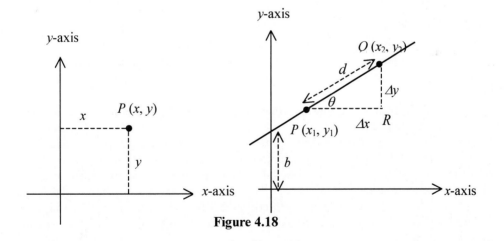

Figure 4.18

Distance between Two Points and the Equation of a Straight Line

Let (x_1, y_1) and (x_2, y_2) be the Cartesian coordinates of two points P and Q, respectively. The distance d between the points P and Q is the hypotenuse of the right triangle PQR in Figure 4.18 and can be obtained by using the Pythagorean theorem. Thus,

$$d = \sqrt{\Delta x^2 + \Delta y^2} = \sqrt{(x_2 - x_1)^2 + (y_2 - y_1)^2} \; .$$

Here,

$$\Delta x = x_2 - x_1 \qquad \text{and} \qquad \Delta y = y_2 - y_1.$$

The line joining P and Q is called the *Cartesian graph*, and the equation of the straight line joining the two points P and Q is given by

$$y = mx + b,$$

where $m = \Delta y / \Delta x = \tan \theta$ (θ is the angle made by the line with the positive x-axis) and b is the length of the intercept of the line with the y-axis at the origin ($x = 0$). The steepness of the line is measured by the ratio m and is called the *slope* of the straight line. If the line passes through the origin, then $b = 0$, and the equation of the line is $y = mx$.

When a relationship between two variables can be written as $y = mx + b$, the relationship is called *linear*. A linear relationship between physical variables is common in physics. For example, when heated, a solid increases its length, and the relationship between the length L and temperature rise ΔT is approximately linear. The relationship is represented by $L = L_0 + L_0 \alpha \Delta T$, where L_0 is the constant initial length of the solid and α is a constant for the solid, called the *coefficient of linear expansion*. Comparing this relation with $y = mx + b$ reveals that here $x = \Delta T$, $y = L$, $m = L_0 \alpha$, and $b = L_0$.

An object can move in any direction in space. That is, it can move left or right, front or back, and up or down. To locate a point in space, three coordinates are needed. In a rectangular Cartesian system, the three axes labeled as x, y, and z are mutually perpendicular, and they intersect at one point: the origin O [Figure 4.19]. The position of a point P is labeled by three numbers (x, y, z), where x is the perpendicular of the point from the y-z plane, y is the distance from the x-z plane, and z is the distance from x-y plane. If (x_1, y_1, z_1) and (x_2, y_2, z_2) are the Cartesian coordinates of two points, say, P and Q, respectively, then the distance between the points is given by

$$d = \sqrt{\Delta x^2 + \Delta y^2 + \Delta z^2} = \sqrt{(x_2 - x_1)^2 + (y_2 - y_1)^2 + (z_2 - z_1)^2} \; .$$

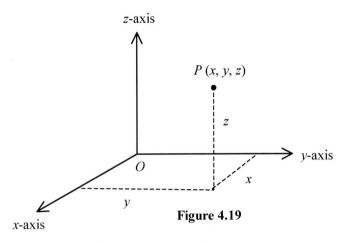

Figure 4.19

Here,

$$\Delta x = x_2 - x_1, \qquad \Delta y = y_2 - y_1, \qquad \text{and} \qquad \Delta z = z_2 - z_1.$$

The idea of the Cartesian coordinate system has been extended to include more than three dimensions. For example, in the special theory of relativity, time becomes the fourth dimension and the time axis is considered mutually perpendicular to the x-, y-, and z-axes. Four-dimensional geometry is used in the theory of relativity.

Example 4.4.1
Determine the slope of a line that passes through the points $(-2, 0)$ and $(4, -4)$. Also, determine the angle θ that the line makes with the positive x axis.

Use the expression for the slope to solve this problem.

$$\text{Slope, } m = \frac{\Delta y}{\Delta x} = \frac{y_2 - y_1}{x_2 - x_1}.$$

Substitute the values

$x_1 = -2, y_1 = 0, x_2 = 4, y_2 = -4$

$$= \frac{-4 - 0}{4 - (-2)} = \frac{-4}{6} = -0.667$$

or $\qquad$ $\tan \theta = -0.667$ $\qquad$ [since $m = \tan \theta$].

Take *inv tan* to get θ $\qquad$ $\theta = \tan^{-1}(-0.667) = -33.7°$.

Example 4.4.2

Determine the equation of a straight line whose slope is 2.0 and that intersects the y-axis at a distance of -1.00 unit. Make a Cartesian plot of the line.

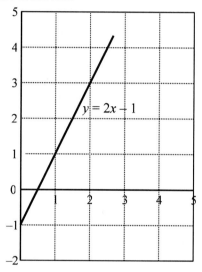

In Cartesian coordinates, the equation of a line is $y = mx + b$.

In this example, $m = 2.0$ and $b = -1.0$.

Thus, the equation of the line is $y = mx + b = 2x - 1$.

The plot of the equation is shown at the right.

Example 4.4.3

Determine the magnitude of the displacement of a person whose initial and final locations in the Cartesian system are (2.0 m, 4.0 m) and (4.0 m, 6.0 m).

The magnitude of the displacement is the straight-line distance between the final and initial locations.

Thus, the magnitude of the displacement $\quad = \sqrt{(x_2 - x_1)^2 + (y_2 - y_1)^2}$

Substitute the values $\qquad = \sqrt{(4.0\,\text{m} - 2.0\,\text{m})^2 + (6.0\,\text{m} - 4.0\,\text{m})^2}$

$\qquad = 2.8\ \text{m}^2.$

4.5   The Polar Coordinate System

The Cartesian coordinate system is most commonly used in introductory physics. However, is not the most convenient system in all situations. In some situations, the polar coordinate system is more convenient. In the polar coordinate system, the coordinates of a point P are (r, θ), as shown in Figure 4.20. In this case, r is the radial distance of the point P from the origin O, and θ is the angle measured counterclockwise from the x-axis to r.

If (x, y) are the Cartesian coordinates of the point P [Figure 4.20], then*

$x = r \cos \theta$

$y = r \sin \theta$.

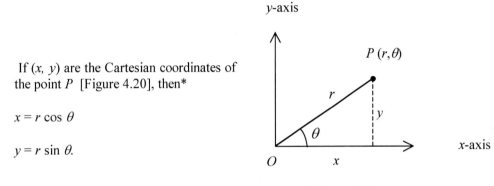

Figure 4.20

The polar coordintes r and θ can also be expressed in terms of x and y. Squaring and adding the preceding equations for x and y yields

$$x^2 + y^2 = r^2\cos^2\theta + r^2\sin^2\theta = r^2 (\cos^2\theta + \sin^2\theta) = r^2$$

or $r = \sqrt{x^2 + y^2}.$

Also, dividing the expressions for x and y, we get,

$$\frac{y}{x} = \frac{r \sin \theta}{r \cos \theta} = \frac{\sin \theta}{\cos \theta} = \tan \theta$$

or $\theta = \tan^{-1}\left(\dfrac{y}{x}\right).$

[* Chapter 5 defines sin, cos, and tan trigonometric functions and gives details about these trigonometric functions and their properties.]

Example 4.5.1
The Cartesian coordinates of a point are (4, 2). Determine the coordinates in the polar system.

In this example, $x = 4$ units and $y = 2$ units. The polar coordinates are (r, θ). Use the preceding two equations to determine r and θ. We have

$$r = \sqrt{x^2 + y^2} = \sqrt{4^2 + 2^2} = \sqrt{20} = 4.5 \text{ units}$$

$$\theta = \tan^{-1}\left(\frac{y}{x}\right) = \tan^{-1}\left(\frac{2}{4}\right) = 0.46 \text{ radian or } 27°.$$

4.6   Choosing a Coordinate System

The choice of a coordinate system is important to solve problems in physics. The choice may determine whether a problem can be solved easily or whether it will be difficult (or impossible) to solve.

In most situations in introductory physics, the Cartesian coordinate system is most convenient. In some situations, such as when there is *radial symmetry*, the polar coordinate system may be preferred.

Note that the origin of the coordinate system can be assumed to be at any point at your convenience. Think about what should be the logical location for the origin. In many situations, the initial location of the object of interest will be the logical choice of the origin. For motion of objects in the vertical direction, a good choice of the origin will be on the ground in most situations. For two-dimensional problems, such as the problems involving the motion of a projectile, use the horizontal direction for the x-axis and the vertical direction for the y-axis, and adopt a sign convention for x and y.

Most physics texts follow the sign convention that the *motion to the right* along a horizontal line is *positive* and motion in the *upward direction* is *positive*, as shown in Figure 4.21. Once a sign convention is selected, stick with it it for the entire problem.

Note that the choice of the *x*-axis to be the horizontal direction and the *y*-axis to be the vertical direction may *not always* be the right choice.

For the motion of an object on an inclined surface, it is more convenient to choose the *x*-axis to be downhill along the inclined surface (with the downhill direction as the positive direction) and the *y*-axis to be perpendicular to the inclined surface (with the positive direction away from the surface), as shown in Figure 4.22. For this choice of axes, the motion of the object along the surface is restricted only to the *x* direction, thus making the solution simpler.

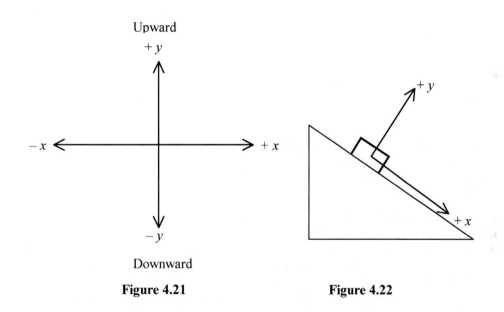

Figure 4.21   **Figure 4.22**

Note that the result of a problem or the physics of a problem does not depend on the choice of the coordinate system nor the sign convention. Your result should be the same, regardless of the choice of axes and the choice of sign convention.

Example 4.6.1

A skier is accelerating down a 30° hill at 2.0 m/s^2. She starts from rest and takes 20.0 seconds to reach the bottom of the hill. Determine the length of the incline.

In this case, the most convenient coordinate system is the Cartesian coordinate system, with the x-axis along the inclined surface and downhill as the positive direction. Choose the origin to be the initial location of the skier. Thus, as the skier accelerates downhill, she moves along the positive x-axis starting from the origin. (See diagram at right.)

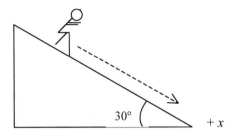

Here, x_0 (the initial location of the skier) = 0.0 m.

v_0 = initial velocity = 0.0 m/s,

a = 2.0 m/s^2,

t = 20.0 s,

x = ?

To solve this problem, we apply the kinematic equation of physics that applies to the x direction when the acceleration is constant. The equation is

$$(x - x_0) = v_0 t + \frac{1}{2} a t^2.$$

Substitute these values in the preceding equation to solve for x, the length of the incline.

$$x - 0 \quad = 0 \cdot 20.0 + \frac{1}{2}(2.0 \text{ m/s}^2)(20 \text{ s})^2,$$

$$x = 400.0 \text{ m}.$$

Conic Sections

4.7   Conic Sections

The second-degree equations representing *circles*, *ellipses*, *parabolas*, and *hyperbolas* are common in physics. For example, the motion of a satellite around the earth is a circle, the path of motion of a planet around the sun is an ellipse, the path of a projectile (such as the motion of a soccer ball in air) in the absence of air resistance is a parabola, and the curve representing the pressure and volume of an ideal gas when its temperature is constant is a hyperbola. These figures were fully investigated in the seventeenth century by Rene Descartes (1596–1650) and Blaise Pascal (1623–1662). The circles, ellipses, parabolas, and hyperbolas are collectively called *conic sections*, because they can be created from the intersection of a plane with a cone, as shown in Figure 4.23. These curves are also defined geometrically and are distinguished by the algebraic equations that represent them.

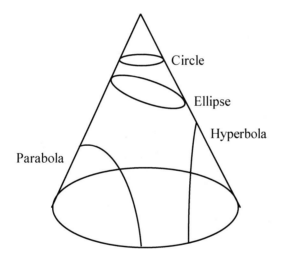

Figure 4.23

Circle

   A *circle* is the simplest conic section. It is defined as the set of all points in the plane equidistant from a fixed point, called the *center* of the circle. The constant distance a is called the *radius* of the circle. If (x_0, y_0) are the coordinates of the center of a circle [Figure 4.24], then the equation of a circle is

$$(x - x_0)^2 + (y - y_0)^2 = a^2.$$

If the center of the circle is at the origin, the equation of a circle reduces to,

$$x^2 + y^2 = a^2.$$

A circle is symmetric about all axes. A *tangent* line drawn at any point to a circle is always *perpendicular* to the *radius* of the circle. Circles have many practical applications, such as circular wheels and gears, merry-go-rounds, ferris wheels, and pizza cutters.

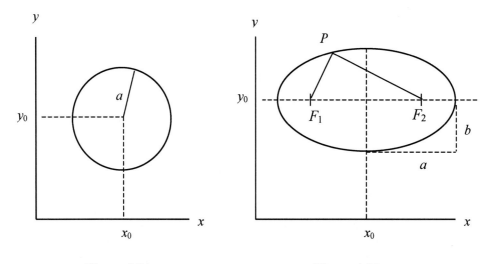

Figure 4.24   **Figure 4.25**

Ellipse

An *ellipse* is the set of all points in the plane for which the sum of the distances from two fixed points is constant. Each of the two points is called a focus (foci are shown as F_1 and F_2 in Figure 4.25) of the ellipse. Thus, for all points P on the ellipse, the sum of the distances $(F_1P + F_2P)$ is constant. An ellipse is a flattened circle that has two centers or two foci.

The line through the two foci that connects each far edge of an ellipse is called the *major axis*. The perpendicular bisector of the major axis is called the *minor axis*. The midway point between the foci is the *center* of the ellipse. The major and minor axes meet at the center. The semimajor axis is the distance from the center through one focus to the ellipse. In Figure 4.25, the semimajor axis is labeled *a*. The semiminor axis is one-half of the minor axis and is labeled *b*.

If (x_0, y_0) are the coordinates of the center [Figure 4.25], then the equation of an ellipse with major axis of length 2*a* and minor axis of length 2*b* *(a > b)* is

$$\frac{(x - x_0)^2}{a^2} + \frac{(y - y_0)^2}{b^2} = 1 \quad \text{with the major axis parallel to the } x\text{-axis.}$$

If the center of an ellipse is at the origin, the equation of the ellipse reduces to

$$\frac{x^2}{a^2} + \frac{y^2}{b^2} = 1.$$

If the major axis is parallel to the *y* axis, the equation is

$$\frac{(x - x_0)^2}{b^2} + \frac{(y - y_0)^2}{a^2} = 1, \text{ or}$$

$$\frac{x^2}{b^2} + \frac{y^2}{a^2} = 1 \text{ if the center is at the origin.}$$

An ellipse is symmetrical about its major axis and its minor axis. Optical and acoustical properties of ellipses are useful in architecture and engineering. You will find that many arches are portions of an ellipse because of its pleasant shape. Gears are cut into elliptic shapes that give nonuniform motion. As mentioned earlier, planets and comets move in elliptic orbits.

Note that circles and ellipses occur frequently in physics problems. These figures can be drawn by scribing a line inside a loop of string stretched out around a fixed single point (center for the circle) or two fixed points (two foci for the ellipse). Note also that a *tangent* line drawn at any point to a circle is *perpendicular* to the *radius* of the circle.

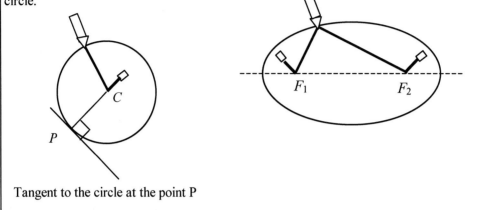

Tangent to the circle at the point P

Parabola

A *parabola* is the set of all points in the plane for which the distance from a fixed point, called the *focus*, and a fixed line, called the *directrix*, are the same. If F is the focus and PQ is the length of the perpendicular distance from a point P on the parabola on the directrix, then $FP = PQ = $ constant [Figure 4.26].

The line that passes through the focus and is perpendicular to the directrix is called the *parabolic axis*. A parabola is symmetric about its parabolic axis. The axis meets the parabola at its *vertex*.

If the vertex of a parabola is at (x_0, y_0), then the equation of a parabola that opens right or left (depending on whether a is positive or negative) is

$$x - x_0 = a\,(y - y_0)^2$$

For a parabola that opens up or down (depending on whether a is positive or negative), the equation is

$$y - y_0 = a\,(x - x_0)^2.$$

A function of the form $y = ax^2 + bx + c$ or $x = ay^2 + by + c$ can be written in one of the preceding two forms. Thus, they represent the parabola in general.

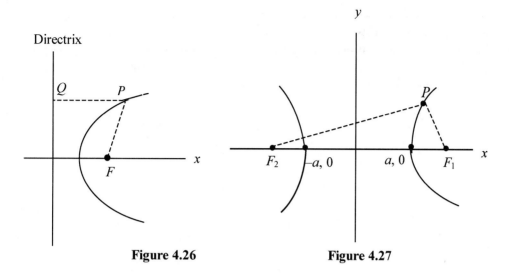

Figure 4.26   **Figure 4.27**

In architecture, you will find many arches that are parabolic in shape, because that shape provides strength. The cables that support suspension bridges hang in the form of a parabola. When a parabola is rotated about it axis it provides a dish-shaped surface called a *paraboloid*. Paraboloids are used as flashlight and headlight reflectors, since, if a source of light (or sound) is placed at the focus of a paraboloid, it reflects outward in parallel paths. Parabolic surfaces are used as antennas, because the surface concentrates the incoming signal at the focus. Parabolic mirrors use solar energy to concentrate light at a point to generate heat.

Hyperbola

A *hyperbola* is the set of all points for which the difference between distances from two fixed points is a constant. Each of the two points is called a *focus*. Thus, for all points P on the hyperbola whose two foci are at F_1 and F_2, $F_1P - F_2P =$ constant [Figure 4.27]. Midway between the foci is the *center* of the hyperbola. The line joining the two foci is called the transverse axis (*x*-axis) and the perpendicular bisector of that line is called the conjugate axis (*y*-axis). If (x_0, y_0) are the coordinates of the center, then the equation of a hyperbola that opens left and right is

$$\frac{(x - x_0)^2}{a^2} - \frac{(y - y_0)^2}{b^2} = 1.$$

The two vertices of the hyperbola are at $(x_0 + a, y_0)$ and $(x_0 - a, y_0)$.

If (x_0, y_0) are the coordinates of the center, then the equation of a hyperbola that opens up and down is

$$\frac{(y - y_0)^2}{a^2} - \frac{(x - x_0)^2}{b^2} = 1.$$

The two vertices are at $(x_0, y_0 + a)$ and $(x_0, y_0 - a)$.

If the center is at the origin, the preceding two equations become

$$\frac{x^2}{a^2} - \frac{y^2}{b^2} = 1 \text{ and } \frac{y^2}{a^2} - \frac{x^2}{b^2} = 1, \text{ respectively.}$$

A hyperbola is symmetric about its *transverse* axis and its *conjugate* axis. A hyperbola forms the basis of a navigation system called Long Range Navigation, or LORAN.

There is a special type of hyperbola (centered at the origin) that does not intersect the x-axis or the y-axis. The x- and y-axes are the *asymptotes* of these hyperbolas. These hyperbolas can be expressed by the equation $xy = k$, where $k \neq 0$ and they are common in physics. For example, the relationship between the pressure P and volume V of an ideal gas when its temperature T is constant can be expressed by the equation $PV = $ constant. If V is plotted along the x- axis and its P along the y-axis, the graph is a hyperbola with the P- and V-axes as its two asymptotes.

Example 4.7.1

Write down the equation $x^2 - 9y^2 + 2x + 36y = 44$ in standard form, and determine what kind of conic section it represents.

$$x^2 - 9y^2 + 2x + 36y = 44$$

First rearrange the terms $\quad x^2 + 2x - 9y^2 + 36y = 44$

Factor out 9 $\quad x^2 + 2x - 9(y^2 - 4y) = 44$

Complete the squares on x and y $(x^2 + 2x + 1) - 9(y^2 - 4y + 4) = 44 + 1 - 3$

$$(x + 1)^2 - 9(y - 2)^2 = 9$$

Divide by 9 on both sides $$\frac{(x+1)^2}{9} - \frac{(y-2)^2}{1} = 1.$$

Comparing this equation with the standard equations of a conic section, we see that the former represents a hyperbola with its center at $(-1, 2)$.

Example 4.7.2

In the Rutherford alpha particle scattering experiment that led to the discovery of the nucleus of an atom, alpha particles were shot toward a thin sheet of gold. An incoming alpha particle was repelled by the nucleus of an atom. If the nucleus was at the origin and the alpha particle traveled along a hyperbolic path whose equation was $4x^2 - y^2 = 64$, how close did the particle come to the nucleus?

The closest point of approach is the vertex of the hyperbola.

$$4x^2 - y^2 = 64$$

Divide by 64 $$\frac{4x^2}{64} - \frac{y^2}{64} = 1$$

$$\frac{x^2}{16} - \frac{y^2}{64} = 1$$

$$\frac{x^2}{4^2} - \frac{y^2}{8^2} = 1.$$

Hence, $a = 4$, and the vertex of the hyperbola is at $(4, 0)$. The closest point of approach is then 4 units from the nucleus.

Exercises

Write down the following equations in standard form and determine what kind of conic sections are represented:

7. $x^2 - 16y^2 + 10x + 9 = 0$

8. $4x^2 + 9y^2 - 16x - 18y - 11 = 0$

9. $x^2 + y^2 - 4x - 8y + 16 = 0$

10. $2x^2 - 4x - y + 4 = 0$

4.8   Parametric Equations

So far, we have described equations in which the variable y has been expressed in terms of x. There are many occasions when it is desirable to express the relationship between two variables in terms of a third variable, upon which the original two variables depend. The equations involving the third variable are called *parametric equations*. For example, the relationship between x and y for a circle passing through the origin is $x^2 + y^2 = a^2$. This relationship can be expressed in terms of a third variable t as, $x = a \cos t$, $y = a \sin t$.

Parametric equations are used in kinematics, trigonometry, and calculus. The following parametric equations used in kinematics are derived in physics books for the horizontal displacement x and the vertical displacement y of an object starting from the origin and moving in a gravitational field:

$$x = x_0 + v_{x,0}t,$$

$$y = y_0 + v_{y,0}t - \frac{1}{2}gt^2.$$

If x_0 and $y_0 = 0$, then these equations reduce to

$$x = v_{x,0}t,$$

$$y = v_{y,0}t - \frac{1}{2}gt^2.$$

where $v_{x,0}$, $v_{y,0}$, and g are constants. The relations are used to determine the trajectory of the object. From the first equation,

$$t = \frac{x}{v_{x,0}},$$

$$y = v_{y,0}t - \frac{1}{2}gt^2$$

$$= v_{y0}\left(\dfrac{x}{v_{x,0}}\right) - \dfrac{1}{2}g\left(\dfrac{x}{v_{x,0}}\right)^2 = \left(\dfrac{v_{y,0}}{v_{x,0}}\right)x - \left(\dfrac{g}{2v_{x,0}{}^2}\right)x^2 .$$

Clearly, this equation is of the type $y = Ax + Bx^2$, so the trajectory is a parabola.

Exercises

10. Show that the equations $x = at$ and $y = b\sqrt{1-t^2}$ parameterize an ellipse and the equations $x = at$ and $y = b\sqrt{1+t^2}$ parameterize a hyperbola.

11. The kinematic equations involving velocity (v) and time (t) as well as distance (x) and time, are given by $v = v_0 + at$ and $x = x_0 + v_0t + 1/2at^2$, respectively (x_0, v_0 and a are constants). In these equations, v and x are separately expressed in terms of t. Show that the direct relationship between v and x is given by $v^2 = v_0{}^2 + 2a(x - x_0)$.

*4.9 Symmetry in Physics

An object or a system is said to be *symmetric* if it appears to be the same under certain operations. The operations include *translation*, *rotation*, and *reflection*. A circle has infinite symmetry both in translation and rotation. When you move a circle in a straight line, regardless of where you move it, the circle still has the same appearance. Also, regardless of how you rotate a circle, it appears exactly the same for an infinite number of orientations, as shown in Figure 4.28. A square has a fourfold rotational symmetry. Its view will be invariant with respect to a rotation of 90°. An equilateral triangle has a threefold rotational symmetry, since it looks the same from any one of the three corners—that is, with respect to a rotation of 120°. A parabola has a *mirror symmetry*, because if you place a mirror perpendicularly on its parabolic axis, you will see its other half in the mirror.

Knowledge of symmetry is important in physics. Take advantage of symmetry when you solve physics problems. For example, there is a symmetry for the motion of an object thrown vertically upward with an initial speed. The object slows down at a constant rate as it moves up, stops momentarily at its maximum height, and then speeds up at the same rate as it falls down. *Time symmetry* exists for this motion, in the sense that the time required for the object to reach its maximum height is equal to the time for its return to its starting point. There is also a *symmetry in the speed* of the object, in the sense that at any distance above the point of release, the speed of the object during the upward trip is equal to the speed at the same point during the downward trip [Figure 4.29 a]. A similar symmetry exists for the motion of a projectile that has been shot at an angle other than 90° or 0°. In this case, the trajectory is a parabola, but the time to reach the maximum height is one-half of the total time of flight, and at any height, the vertical components of velocity are equal and opposite [4.29 b].

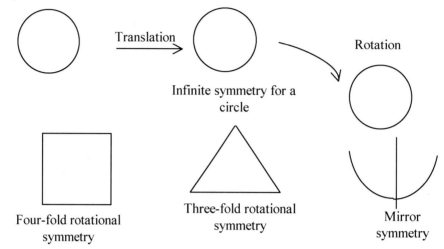

Figure 4.28

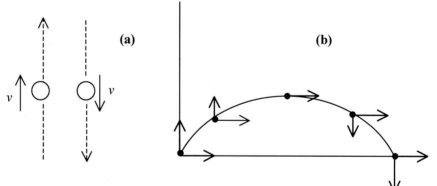

Figure 4.29 : Symmetry in speed for (a) vertical motion and (b) projectile motion

The universe has different symmetries. It is interesting and important to know that the symmetries of the universe are related to the fundamental conservation principles (conservation of linear momentum, conservation of angular momentum, and conservation of energy) of physics. You will learn about these conservation principles from your physics course. In fact, *for any conservation principle, there exists a type of symmetry in nature and vice versa.*

*4.10 Non-Euclidean Geometry

The geometry that's been discussed so far is called *plane geometry* or *Euclidean geometry*. In this geometry, the shortest distance between two points is a straight line, and the sum of the three angles in a triangle is 180°. But suppose you need to know the shortest flight path between two airports that are very far apart.

Since the earth is nearly a sphere, this flight path would be different from the one gotten by using a ruler to draw a straight line on a flat map joining the two airports on paper.

The geometry on a flat surface is Euclid's; geometry on a curved surface (such as the surface of the earth) is non-Euclidean. To determine the path of an aircraft or the distance between two cities, non-Euclidean geometry is relevant.

In the language of geometry, the two dimensional surface of a sphere (such as the earth) is curved, and the surface has *positive curvature*. The sum of three angles of a triangle drawn connecting three points on the surface of the sphere is *more* than 180°. For example, start at the North Pole and move to the equator. Turn 90° and move along the equator a quarter of the way around the earth. Then, turn 90° to go to the pole. You will end up 90° from your original path. The triangle has three angles of 90° each. So the sum of the angles is 270°.

The two-dimensional surface on a flat piece of paper has *zero curvature*, and the sum of the three angles of a triangle on the paper is 180°. A saddle-shaped surface is also curved, but it has *negative curvature*.

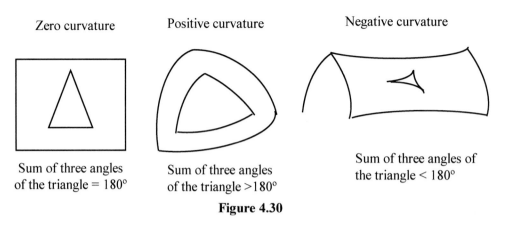

Zero curvature

Positive curvature

Negative curvature

Sum of three angles
of the triangle = 180°

Sum of three angles
of the triangle >180°

Sum of three angles of
the triangle < 180°

Figure 4.30

The sum of the three angles of a triangle on a surface of negative curvature is *less* than 180°. Figure 4.30 shows surfaces with zero, positive, and negative curvatures and also shows what a triangle drawn on these surfaces looks. On a curved surface, the lines forming straight lines are not all straight from a three-dimensional point of view, but they are the shortest distances between points if we are confined to the curved surface.

Non-Euclidean geometry is useful when a surface is curved. For example, in the general theory of relativity, which is a new interpretation of gravitation, gravity is related to the *geometry of space–time*, and space–time is *curved* when gravity is strong. Non-Euclidean geometry is useful in studying the general theory of relativity. Because space–time is curved near the sun owing to its own gravity, light beams bend when passing near the sun. You will learn more about this phenomenon from your introductory physics course.

Answers to Exercises in Chapter 4

1. 3.0×10^{-4} rad, 4 cm

2. $A = 30°$, $B = 115°$, $C = 55°$, $D = 60°$

3. $\psi = 75°$, $\phi = 75°$, $v = 60°$, $RS = 2.8$, $RT = 3.3$

4. 5490 kg/m^3

5. 0.21 cm^2

6. 7.3 cm^3

7. $(x + 5)^2/16 - y^2 = 1$, hyperbola

8. $(x - 2)^2/9 + (y - 1)^2/4 = 1$, ellipse

9. $(x - 2)^2 + (y - 4)^2 = 4$, circle

10. $y = 2(x - 1)^2 + 2$, parabola

Chapter 5
Trigonometry: A Powerful Tool for Solving Real-World Problems

Trigonometry deals with the interrelationships among the sides and angles of triangles. It is based primarily on the special properties of a right triangle. It is a powerful tool for scientists and engineers for taking measurements as well as for solving real-world problems.

5.1   Angles in Trigonometry

In trigonometry, *angles* are created because of the *rotation*, in a plane, of a straight line about a point O on the line, called the *origin*.   If the line OA is rotated in a plane about O onto OB, an angle AOB is generated [Figure 5.1]. The angle is considered to be *positive* or *negative* depending on whether the rotation is *counterclockwise* or *clockwise*.   As mentioned in Chapter 4, angles can be measured in degrees or in radians.

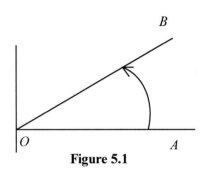

Figure 5.1

$$360° = 2\pi \text{ radians,}$$

$$1 \text{ radian} = 180/\pi \text{ degrees} = 57.3° \qquad 1° = \pi/180 \text{ radians}$$

An angle is considered to be in a standard position if its vertex is at the origin and its initial side is on the positive *x*-axis of a rectangular Cartesian coordinate system. An angle is considered to be a first-, second-, third-, or fourth-quadrant angle, depending on whether its terminal side falls in the first, second, third, or fourth quadrant, respectively, of the coordinate system [Figure 5.2].

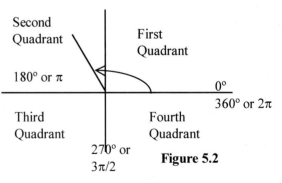

Figure 5.2

Right-Angle Trigonometry

5.2 Right Triangles and the Pythagorean Theorem

A *right triangle* is a triangle in which one angle is 90°. The lengths of the base b, height a, and hypotenuse c, of a right triangle [Figure 5.3] are related by

$$c^2 = a^2 + b^2.$$

This is known as the *Pythogorean theorem*, a mathematical theorem that is very useful for finding the length of one side of a right triangle when the other two sides are known.

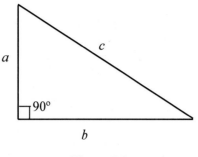

Figure 5.3

In physics, the Pythogorean theorem is useful in many practical situations. An example in kinematics is to determine the displacement of an object, as illustrated in the following example.

Example 5.2.1

The amount of displacement of an object is defined as the shortest distance between its final location and its initial location. A person travels 3 mi east and then travels 4 mi north. What is the amount of displacement of the person from his initial location?

The person has traveled along the two perpendicular sides of a right triangle. The sketch of the problem is shown to the right. The hypotenuse of the triangle is the required displacement.
Thus, in this case,
$b = 3$ mi and $a = 4$ mi.
From the Pythagorean theorem,
the magnitude of the displacement
is given by $c^2 = a^2 + b^2.$
$\qquad = (3 \text{ mi})^2 + (4\text{mi})^2$
$\qquad = 9 \text{ mi}^2 + 16 \text{ mi}^2 = 25 \text{ mi}^2$
The displacement is, then, $c = \sqrt{25}$ mi $= 5$ mi.

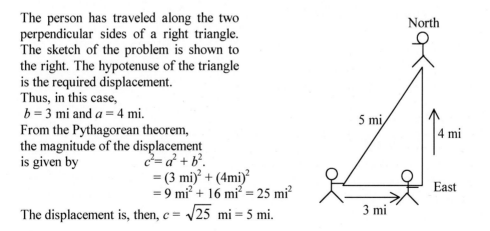

The result makes sense because the hypotenuse of a right triangle is always greater than the other two sides.

5.3   Trigonometric Functions

Three *trigonometric functions* are often used in physics. These are defined in terms of the ratio of two sides of a right triangle. The functions are called *sine* (or *sin*), *cosine* (or *cos*), and *tangent* (or *tan*).   Consider a right triangle whose one acute angle is θ at one vertex as shown in Figure 5.4. Its *hypotenuse* is h, the side *opposite* θ is h_o (also called *rise*), and the side *adjacent* to θ is h_a (also called *run*.)   Then, *sin*, *cos*, and *tan* are defined by the following ratios:

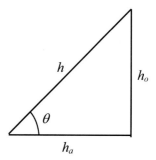

Figure 5.4

$$\sin \theta = \frac{h_o}{h} = \frac{opposite}{hypotenuse} \text{ or } \frac{rise}{hypotenuse}$$

$$= \frac{h_o}{\sqrt{h_o^2 + h_a^2}} \text{ (Since from Pythagorean theorem } h^2 = h_a{}^2 + h_o{}^2\text{)}$$

$$\cos \theta = \frac{h_a}{h} = \frac{adjacent}{hypotenuse} \text{ or } \frac{run}{hypotenuse} = \frac{h_a}{\sqrt{h_o^2 + h_a^2}}$$

$$\tan \theta = \frac{h_o}{h_a} = \frac{opposite}{adjacent} \text{ or } \frac{rise}{run}.$$

The other three trigonometric functions are *cosecant* (or *csc*), *secant* (or *sec*), and *cotangent* (or *cot*). These are defined as follows:

$$\csc \theta = \frac{h}{h_o} = \frac{1}{\sin \theta}$$

$$\sec \theta = \frac{h}{h_a} = \frac{1}{\cos\theta}$$

$$\cot \theta = \frac{h_a}{h_o} = \frac{1}{\tan \theta}$$

Now, $\tan \theta = \dfrac{h_o}{h_a}$. This can be written as

$$\tan \theta = \frac{h_o}{h_a} = \frac{h_o/h}{h_a/h} \quad \text{[by dividing the numerator and denominator by } h].$$

From the definitions of *sin* and *cos*, it is seen that the right-hand side of the preceding equation is the ratio of *sin* and *cos*. Therefore,

$$\tan\theta = \frac{\sin \theta}{\cos\theta}.$$

Similarly, we can show that

$$\cot \theta = \frac{\cos\theta}{\sin \theta}.$$

In most situations in physics, you need to know only *sin*, *cos*, and *tan*. An easy way to remember how they are defined is to remember the mnemonic: *soh cah toa* (usually pronounced as "so-kah-toe-ah.") This means that

$$\textbf{soh} = \sin = \frac{opposite}{hypotenuse}.$$

$$\textbf{cah} = \cos = \frac{adjacent}{hypotenuse}.$$

$$\textbf{toa} = \tan = \frac{opposite}{adjacent}.$$

The values of *sin*, *cos*, and *tan* for commonly used angles are shown in Table 5.1.

Table 5.1: Values of *sin*, *cos*, and *tan* for commonly used angles

θ	$\sin \theta$	$\cos \theta$	$\tan \theta$
0°	0	1	0
30°	$\frac{1}{2} = 0.5$	$\sqrt{3}/2 = 0.87$	$1/\sqrt{3} = 0.58$
45°	$1/\sqrt{2} = 0.71$	$1/\sqrt{2} = 0.71$	1
60°	$\sqrt{3}/2 = 0.87$	$\frac{1}{2} = 0.5$	$\sqrt{3} = 1.73$
90°	1	1	∞

Figure 5.5 shows the signs (+ or –) that *sin*, *cos*, and *tan* take on for angle θ in different quadrants (0 to 2π or 360°). The angles are measured *counterclockwise* from the *x*-axis, as shown in the figure. The signs are determined by the sign of *x* and *y* in different quadrants. The negative angles are measured below the *x*-axis. This means that –60° is 360° – 60° = 300°.

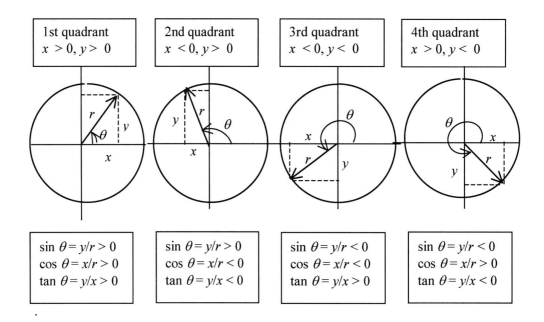

Figure 5.5

This is usually remembered as *all, sin, tan, cos*, meaning *all* are positive in the *first* quadrant, *sin* is positive in the second quadrant, *tan* is positive in the third quadrant, and *cos* is positive in the fourth quadrant.

The signs of *sin, cos,* and *tan* in four quadrants are also shown in Table 5.2.

The plots of the *sin, cos,* and *tan* for –2π to +2π are shown in Figure 5.6.

y = sin x   y = cos x

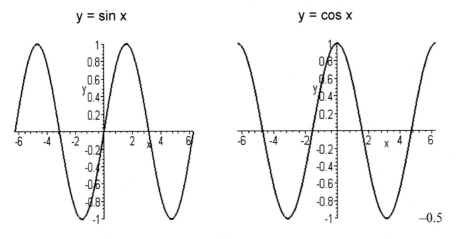

y = tan x

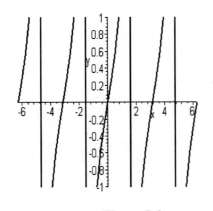

Figure 5.6

Table 5.2: Sign of Trigonometric Functions in Four Quadrants

Quadrant	*sin*	*cos*	*tan*
I	+	+	+
II	+	−	−
III	−	−	+
IV	−	+	+

The plots also show that the signs of the trigonometric functions can be positive and negative. The variations of the values of *sin*, *cos*, and *tan*, as seen from their plots, are summarized in Table 5.3.

Table 5.3: Variations of *sin*, *cos*, and *tan* in Four Quadrants

Quadrant	*sin*	*cos*	*tan*
I	0 to +1	+1 to 0	0 to +∞
II	+1 to 0	0 to −1	−∞ to 0
III	0 to −1	−1 to 0	0 to +∞
IV	−1 to 0	0 to +1	−∞ to 0

> **Note:** In general,
>
> $-1 \leq \sin \theta \leq +1$
>
> $-1 \leq \cos \theta \leq +1$
>
> $-\infty \leq \tan \theta \leq \infty$

Example 5.3.1
Find the height of an equilateral triangle whose sides are each 10 cm in length.

We can solve the problem by using i) the Pythagorean theorem, as well as ii) a trigonometric function.

   i) The height *h* bisects the base.

   Therefore, from the Pythagorean

   theorem,

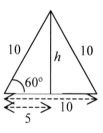

$$5^2 + h^2 = 10^2$$

or $h = \sqrt{(10\,cm)^2 - (5cm)^2} = 8.7$ cm.

ii) For an equilateral triangle, all angles are 60°. Therefore, using the *sin* function, we obtain

$$\sin 60° = \frac{h}{10\,cm} \quad [\text{since } \sin \theta = \frac{opposite}{hypotenuse}].$$

Multiply by 10 $h = (10$ cm$) \sin 60° = (10$ cm$) \times 0.866 = 8.7$ cm.

Example 5.3.2

On a sunny day, a tall pillar casts a shadow that is 6.3 m long. The angle between the sun's rays and the ground is 47°. Determine the height of the pillar.

For this problem, the angle $\theta = 47°$ and

the distance (h_a or the "adjacent") = 6.3

m.

The height (h_o or "opposite") is

unknown.

The trigonometric function that

involves θ, h_a, and h_o is tan θ.

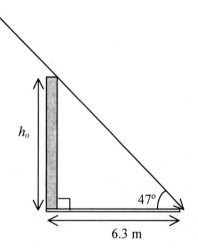

$$\tan \theta \quad = \frac{h_o}{h_a} \text{ or} \frac{opposite}{adjacent}$$

or $h_o = h_a \tan \theta$

Substitute the values $= 6.3$ m $\times \tan (47°)$

$= 6.3$ m $\times 1.07$

$= 6.5$ m.

Therefore, the height of the pillar is 6.5 m.

Example 5.3.3

A box is pulled up a 36° incline. When it has moved 4.0 m along the incline, what distance x has it moved in the horizontal direction and what elevation y has it achieved?

In this case, $\theta = 36°$, h or hypotenuse = 4.0 m

h_o or "opposite" = y (to be determined)

h_a or "adjacent" = x (to be determined)

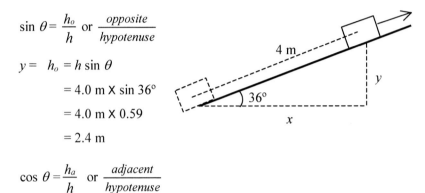

$$\sin \theta = \frac{h_o}{h} \text{ or } \frac{opposite}{hypotenuse}$$

$$y = h_o = h \sin \theta$$

$$= 4.0 \text{ m} \times \sin 36°$$

$$= 4.0 \text{ m} \times 0.59$$

$$= 2.4 \text{ m}$$

$$\cos \theta = \frac{h_a}{h} \text{ or } \frac{adjacent}{hypotenuse}$$

$$x = h_a = h \cos \theta = 4.0 \text{ m} \times \cos 36° = 4.0 \text{ m} \times 0.81 = 3.2 \text{ m}$$

Exercise

Using trigonometric functions, determine the lengths of the sides of the right triangles, which are not given using trigonometric functions.

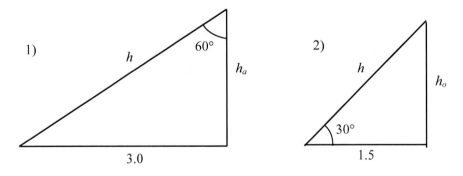

3. In a stairway, each step is set back 25 cm from the next lower step and the stairway rises at an angle of 38° with respect to the horizontal. Determine the height of each step.

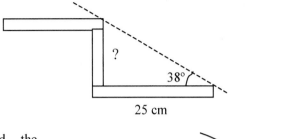

25 cm

4. Where the road is curved, the roadbed is elevated at an angle with respect to the horizontal to avoid skidding when a car follows the curved road. This angle is called a banking angle. If the banking angle is 15° and the outside of the curved road is elevated by 2 m from the inside of the curve, determine the width of the road where it is curved.

5. The gondola ski lift in Colorado is 2830 m long. If the ski lift rises 14.6° from the horizontal, determine the height of the top of the lift relative to the ground.

5.4   Common Trigonometric Identities

From the Pythagorean theorem and using Figure 5.4, we find that

$h^2 = h_a^2 + h_o^2 = (h \sin \theta)^2 + (h \cos \theta)^2$ (from the definitions of *sin* and *cos*).

Canceling h^2 from both sides, we get

$$\sin^2 \theta + \cos^2 \theta = 1.$$

This is a *trigonometric identity*. There are several other trigonometric identities; most of which can easily be proved.

The following trigonometric identities and relations are useful in physics and in solving problems:

Pythagorean relations

$\sin^2\theta + \cos^2\theta = 1$ $\sec^2\theta = 1 + \tan^2\theta$

$\csc^2\theta = 1 + \cot^2\theta$

Reciprocal relations

$\csc\theta = \dfrac{1}{\sin\theta}$ $\sec\theta = \dfrac{1}{\cos\theta}$

$\cot\theta = \dfrac{1}{\tan\theta}$

Quotient relations

$\tan\theta = \dfrac{\sin\theta}{\cos\theta}$ $\cot\theta = \dfrac{\cos\theta}{\sin\theta}$

Example 5.4.1

Show that $\sec\theta - \cos\theta = \tan\theta \cdot \sin\theta$

We have $\sec\theta - \cos\theta.$

Substitute $1/\cos\theta$ for $\sec\theta$ $= 1/\cos\theta - \cos\theta$

$$= \frac{1 - \cos^2\theta}{\cos\theta}$$

Substitute $\sin^2\theta$ for $1 - \cos^2\theta$ $= \dfrac{\sin^2\theta}{\cos\theta}$ $[\sin^2\theta + \cos^2\theta = 1]$

$$= \frac{\sin\theta}{\cos\theta} \cdot \sin\theta.$$

Substitute $\tan\theta$ for $\sin\theta/\cos\theta$ $= \tan\theta \cdot \sin\theta.$

Exercise
Prove the following:

6. $\dfrac{\sin\theta}{1+\cos\theta} = \dfrac{1-\cos\theta}{\sin\theta}$ [*Hint*: Multiply the left-hand side by $(1-\cos\theta)$.]

7. $\dfrac{1-\tan^2\theta}{1+\tan^2\theta} = 1 - 2\sin^2\theta$ [*Hint*: $\tan\theta = \sin\theta/\cos\theta$.]

8. $\dfrac{\tan\theta-\cot\theta}{\tan\theta+\cot\theta} = 1 - 2\cos^2\theta$

9. $\dfrac{\sec\theta}{\tan\theta+\cot\theta} = \sin\theta$

Angle-sum and angle-difference relations

$\sin(A+B) = \sin A \cos B + \cos A \sin B$

$\sin(A-B) = \sin A \cos B - \cos A \sin B$

$\cos(A+B) = \cos A \cos B - \sin A \sin B$

$\cos(A-B) = \cos A \cos B + \sin A \sin B$

$\tan(A+B) = \dfrac{\tan A + \tan B}{1 - \tan A \tan B}$ $\tan(A-B) = \dfrac{\tan A - \tan B}{1 + \tan A \tan B}$

$\cot(A+B) = \dfrac{\cot A \cot B - 1}{\cot B + \cot A}$ $\cot(A-B) = \dfrac{\cot A \cot B + 1}{\cot B - \cot A}$

Example 5.4.2
Show that $\sin 2A = 2\sin A \cos A$.

We have $\sin 2A = \sin(A+A)$.

Use the formula for $\sin(A+B)$ $= \sin A \cos A + \cos A \sin A$

                                       $= 2\sin A \cos A$.

Exercises
Show that the following are true:

10. $\tan(\theta+\pi/4) = \dfrac{\cos\theta+\sin\theta}{\cos\theta-\sin\theta}$ 11. $\sin(A+B)\sin(A-B) = \sin^2 A - \sin^2 B$

12. $\cos(\pi/3 + B) = \sin(\pi/6 - B)$

13. $\cos(A + B)\cos B + \sin(A + B)\sin B = \cos A$

Double-angle relations

$$\sin 2\theta = 2\sin\theta\cos\theta = \frac{2\tan\theta}{1 + \tan^2\theta}$$

$$\cos 2\theta = \cos^2\theta - \sin^2\theta = 2\cos^2\theta - 1 = 1 - 2\sin^2\theta = \frac{1 - \tan^2\theta}{1 + \tan^2\theta}$$

$$\tan 2\theta = \frac{2\tan\theta}{1 - \tan^2\theta} \qquad\qquad \cot 2\theta = \frac{\cot^2\theta - 1}{2\cot\theta}$$

Example 5.4.3
Show that $\cos^4 A - \sin^4 A = \cos 2A$

We have
$$\cos^4 A - \sin^4 A = (\cos^2 A)^2 - (\sin^2 A)^2.$$
$$= (\cos^2 A + \sin^2 A)(\cos^2 A - \sin^2 A)$$

Substitute 1 for $\cos^2 A + \sin^2 A$
$$= 1.(\cos^2 A - \sin^2 A)$$
$$= \cos^2 A - \sin^2 A$$
$$= \cos 2A.$$

Exercises
Show that the following identities are true:

14. $\dfrac{1 + \sin 2\theta}{1 + \cos 2\theta} = \tfrac{1}{2}(1 + \tan\theta)^2$

15. $\dfrac{1 - \cos 2\theta}{1 + \cos 2\theta} = \tan^2\theta$

16. $\cos 2\theta = \dfrac{1 - \tan^2\theta}{1 + \tan^2\theta}$

17. $\dfrac{\sin 2\theta}{\sin\theta} + \dfrac{\cos 2\theta}{\cos\theta} = \dfrac{2\sin 3\theta}{\sin 2\theta}$

Function-sum and function-difference relations

$$\sin A + \sin B = 2 \sin \left(\frac{A + B}{2} \right) \cos \left(\frac{A - B}{2} \right)$$

$$\sin A - \sin B = 2 \sin \left(\frac{A - B}{2} \right) \cos \left(\frac{A + B}{2} \right)$$

$$\cos A + \cos B = 2 \cos \left(\frac{A + B}{2} \right) \cos \left(\frac{A - B}{2} \right)$$

$$\cos A - \cos B = 2 \sin \left(\frac{A + B}{2} \right) \sin \left(\frac{B - A}{2} \right)$$

The following example shows how the function-sum relation is used in describing a physical phenomenon.

Example 5.4.4 (Standing wave)

Two identical waves traveling in opposite directions can be represented by the equations $y_1 = A \sin (kx - \omega t)$ and $y_2 = A \sin (kx + \omega t)$ where A, k, and ω are characteristic constants of the waves (called amplitude, wave number, and angular frequency, respectively) and x and t represent position and time, respectively. Show that the resulting wave $y = y_1 + y_2 = 2A \sin kx \cos \omega t$.

Following is a sketch of the problem:

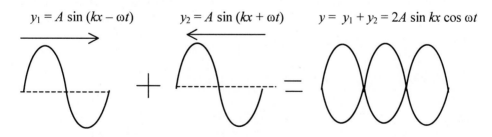

$y_1 = A \sin (kx - \omega t)$ $y_2 = A \sin (kx + \omega t)$ $y = y_1 + y_2 = 2A \sin kx \cos \omega t$

To solve the problem mathematically, use the addition relation for sine functions:

$$y \quad = y_1 + y_2$$

$$= A \sin (kx - \omega t) + A \sin (kx + \omega t)$$

Factor out A

$$= A \left[\sin (kx - \omega t) + \sin (kx + \omega t) \right]$$

Use the addition relation

$$= A \cdot 2 \sin \frac{(kx - \omega t) + (kx + \omega t)}{2} \cos \frac{(kx - \omega t) - (kx + \omega t)}{2}$$

[Since $\sin A + \sin B = 2 \sin (A + B)/2 \cos (A - B)/2$]

$$= 2A \sin \left(\frac{2kx}{2} \right) \cos \left(\frac{-2\omega t}{2} \right).$$

Cancel 2 from *sin* and *cos*

$$= 2A \sin kx \cos (-\omega t)$$

$$= 2A \sin kx \cos \omega t. \quad \text{[since } \cos (-\theta) = \cos \theta, \text{ see the}$$

reduction formulas that follow.]

The resulting wave y is known as a standing wave in physics.

Exercises

18. Two waves traveling in a medium are represented by the equations $y_1 = A \sin (kx - \omega t)$ and $y_2 = A \sin (kx - \omega t + \phi)$ where A, k, and ω are characteristic constants of the waves (called amplitude, wave number, and angular frequency, respectively), ϕ is called the phase difference between the two waves, and x and t represent position and time, respectively.   Show that the resulting wave

$$y = y_1 + y_2 = 2A \cos \frac{\phi}{2} \sin \left(kx - \omega t + \frac{\phi}{2} \right).$$

[The resulting wave shows interference of waves.]

19. Two sinusoidal waves of frequencies f_1 and f_2 are represented at a fixed point in space by $y_1 = \sin 2\pi f_1 t$ and $y_2 = \sin 2\pi f_2 t$. Show that the resultant displacement

$$y = y_1 + y_2 = 2A \cos 2\pi \left(\frac{f_1 - f_2}{2} \right) t \sin 2\pi \left(\frac{f_1 + f_2}{2} \right) t .$$

[The resulting wave shows what is called beats.]

Function-product relations

$$\sin A \cos B = \frac{1}{2}\sin (A + B) + \frac{1}{2}\sin (A - B)$$

$$\cos A \cos B = \frac{1}{2}\cos (A - B) + \frac{1}{2}\cos (A + B)$$

$$\sin A \sin B = \frac{1}{2}\cos (A - B) - \frac{1}{2}\cos (A + B)$$

$$\cos A \sin B = \frac{1}{2}\sin (A + B) - \frac{1}{2}\sin (A - B)$$

Half angle relations

$$\sin \frac{\theta}{2} = \sqrt{\frac{1 - \cos \theta}{2}} \qquad\qquad \tan \frac{\theta}{2} = \sqrt{\frac{1 - \cos \theta}{1 + \cos \theta}}$$

$$\cos \frac{\theta}{2} = \sqrt{\frac{1 + \cos \theta}{2}} \qquad\qquad \cot \frac{\theta}{2} = \sqrt{\frac{1 + \cos \theta}{1 - \cos \theta}}$$

Exercises

Show that the following identities are true:

20. $\cot \dfrac{\theta}{2} = \dfrac{1}{\csc \theta - \cot \theta}$

21. $\tan \dfrac{\theta}{2} = \dfrac{\sin \theta}{1 + \cos \theta}$

Reduction Formulas

$\cos (-\theta) = + \cos \theta$

$\sin (-\theta) = - \sin\theta$

$\tan (-\theta) = - \tan\theta$

$\sin (90° \pm \theta) = + \cos \theta$

$\cos (90° \pm \theta) = \mp \sin \theta$

$\tan (90° \pm \theta) = \mp \cot \theta$

$\sin (180° \pm \theta) = \mp \sin\theta$

$\cos (180° \pm \theta) = - \cos \theta$

$\tan (180° \pm \theta) = \pm \tan \theta$

$\sin (270° \pm \theta) = - \cos\theta$

$\cos (270° \pm \theta) = \pm \sin \theta$

$\tan (270° \pm \theta) = \mp \cot \theta$

$\sin (360° \pm \theta) = \pm \sin\theta$

$\cos (360° \pm \theta) = + \cos \theta$

$\tan (360° \pm \theta) = \pm \tan \theta$

Example 5.4.5

Express the following trigonometric functions in terms of angles in the first quadrant: sin (–60°), sin 150°, cos (–60°), cos 330°, tan 120°, tan 210°.
Then use Table 5.1 for the answer. Do not use a calculator.

$$\sin(-60°) = -\sin(60°) \text{ [since } \sin(-\theta) = -\sin\theta]$$

$$= -0.87$$

$$\sin 150° = \sin(180 - 30)° = \sin 30° \text{ [since } \sin(180° - \theta) = \sin\theta]$$

$$= 0.5$$

$$\cos(-60°) = \cos 60° \text{ [since } \cos(-\theta) = +\cos\theta]$$

$$= 0.5$$

$$\tan 120° = \tan(180 - 60)° = -\tan 60° \text{ [since } \tan(180° - \theta) = -\tan\theta]$$
$$= -1.73$$

$$\tan 210° = \tan(180 + 30)° = \tan 30° \text{ [since } \tan(180° + \theta) = \tan\theta]$$
$$= 0.58$$

5.5   Inverse Trigonometric Functions

Any angle θ whose *sin* is x is denoted by *arcsin x* or $sin^{-1}x$. Thus, if sin $\theta = x$, $\theta = sin^{-1}x$. Similar notation is used to define other inverse trigonometric functions: $cos^{-1}x$, $tan^{-1}x$, $csc^{-1}x$, $sec^{-1}x$, and $cot^{-1}x$.

Note: The notation $sin^{-1}x$ is not the same as $(sin\ x)^{-1}$.

$(sin\ x)^{-1} = 1/sin\ x = csc\ x$. But $sin^{-1}x$ is the inverse *sin x* function.

If any inverse trigonometric function, such as $sin^{-1}x$, is known, it is possible to express all trigonometric functions in terms of x. For example, let $\theta = sin^{-1}x$. Therefore, sin $\theta = x$. You can determine other trigonometric functions in terms of x, as the following shows:

$$\cos\theta = \sqrt{1 - \sin^2\theta} \qquad\qquad (\text{since } \sin^2\theta + \cos^2\theta = 1)$$

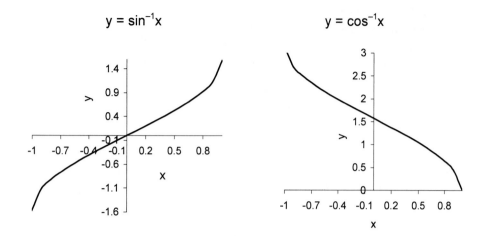

$y = \sin^{-1}x$

$y = \cos^{-1}x$

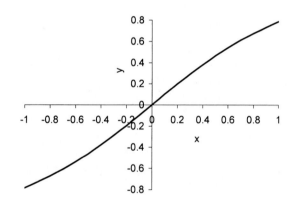

$y = \tan^{-1}x$

Figure 5.7

$$= \sqrt{1-x^2} \qquad\qquad \tan\theta = \frac{\sin\theta}{\cos\theta} = \frac{x}{\sqrt{1-x^2}}$$

$$\csc\theta = \frac{1}{\sin\theta} = \frac{1}{x} \qquad\qquad \sec\theta = \, = \frac{1}{\cos\theta} = \frac{1}{\sqrt{1-x^2}}$$

$$\cot \theta = = \frac{1}{\tan \theta} = \frac{\sqrt{1-x^2}}{x}$$

The plots of $sin^{-1}x$, $cos^{-1}x$, and $tan^{-1}x$ are shown in Figure 5.7.

Fundamental identities involving inverse trigonometric functions are

$$sin^{-1}x + cos^{-1}x = \pi/2$$

and

$$tan^{-1}x + cot^{-1}x = \pi/2.$$

Note: Inverse trigonometric functions can be easily determined by pressing sin^{-1}, cos^{-1}, and tan^{-1} on your calculator. When you use a calculator to determine the value, note that the calculator always displays the smallest correct angle. For example, $sin^{-1}(0.5)$ is both 30° and 150°. But calculator will display only 30°. Similarly, $sin^{-1}(-0.5)$ is −30° and 210° but the calculator will display only −30°. The displayed value *may not* be the right answer in some situations. Use the knowledge of which quadrant you are working in to find which answer is correct.

Example 5.5.1

A lakefront drops off gradually at an angle θ, as shown. A lifeguard rows out straight 20.0 m and drops a weighted fishing line to find the depth at that distance. He found the depth to be 1.28 m. What is the angle θ?

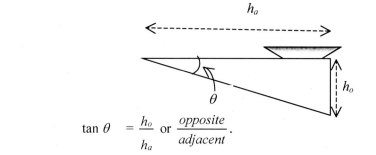

We know that

$$\tan \theta = \frac{h_o}{h_a} \text{ or } \frac{opposite}{adjacent}.$$

Substitute the values	$= \dfrac{1.28\ m}{20\ m}$.

Divide the numbers	$= 0.064$.

Take the *inverse tan* to get the result θ	$= \tan^{-1}(0.064) = 3.7°$.

Exercises

22. Find the values of the following operations:

$\cos^{-1}(0.56)$, $\tan^{-1}(2.5)$, and $\sin^{-1}(0.92)$.

Using inverse trigonometric functions, determine the angle θ of the right triangles:

23)

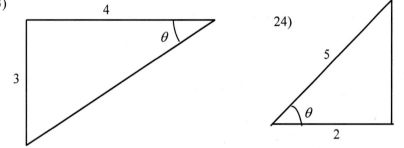

24)

25. If $\theta = \cos^{-1}x$, determine $\sin \theta$, $\cos \theta$, $\tan \theta$, $\cot \theta$, $\csc \theta$, and $\sec \theta$ in terms of x.

26. If $\theta = \tan^{-1}x$, determine $\sin \theta$, $\cos \theta$, $\tan \theta$, $\cot \theta$, $\csc \theta$, and $\sec \theta$ in terms of x.

5.6   Series   Expansion   of   Trigonometric   Functions   and   Small-Angle Approximations

Trigonometric functions and inverse trigonometric functions can be expanded as follows:

$$\sin x = x - \frac{x^3}{3!} + \frac{x^5}{5!} - \ldots \qquad\qquad \sin^{-1} x = x + \frac{1}{6}x^3 + \frac{3}{40}x^5 + \ldots$$

$$\cos x = 1 - \frac{x^2}{2!} + \frac{x^4}{4!} - \ldots \qquad\qquad \cos^{-1} x = \frac{\pi}{2} - \sin^{-1} x$$

$$\tan x = x + \frac{x^3}{3} + \frac{x^5}{15} + \ldots \qquad\qquad \tan^{-1} x = x - \frac{x^3}{3} + \frac{x^5}{5} - \ldots$$

The right-hand side of the preceding expressions contains an infinite number of terms. A polynomial with an infinite number of terms is called a *power series*; therefore, the preceding expansions are called power series expansions of trigonometric or inverse trigonometric functions.

It is often desirable to *approximate* the solution of a problem because the exact solution may be difficult to obtain or may not be necessary. The preceding power series expansions are very helpful for approximations. For example, when x is in *radians* and is very small ($x \ll 1$), then x^2 and other terms are negligibly small and

$$\sin x \approx x$$

$$\cos x \approx 1 - \frac{x^2}{2} \quad \text{(keeping up to the 2}^{\text{nd}}\text{ power of } x\text{)}$$

$$\approx 1 \qquad \text{(keeping only the first term)}$$

$$\tan x \approx x, \text{ etc.}$$

Example 5.6.1
Determine the small-angle approximation for the expression $4\sin^2\theta/\cos\theta$.

$$\frac{4\sin^2\theta}{\cos\theta} \approx \frac{4(\theta)^2}{1} \qquad\qquad [\text{since } \sin\theta \approx \theta \text{ and } \cos\theta \approx 1]$$

$$= 4\theta^2$$

Exercise

30. A small mass attached to a string attached to a fixed vertical support forms a simple pendulum. When the mass is moved to one side from its position of rest by an angle, say, θ, with the vertical line and then released, it moves back and forth. The time the mass takes to complete one full swing is called its period T and is given by

$$T = 2\pi \sqrt{\frac{L}{g}} \, [1 + \tfrac{1}{4} \sin^2 (\theta/2) + 9/64 \sin^4 (\theta/2) + ...],$$

where L is the length of the pendulum (the distance of the string from the fixed support plus the radius of the mass, assuming the mass is a uniform sphere), and g is the local acceleration due to gravity.

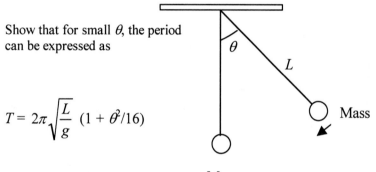

Show that for small θ, the period can be expressed as

$$T = 2\pi \sqrt{\frac{L}{g}} \, (1 + \theta^2/16)$$

Mass

Non-Right-Angle Trigonometry

5.7   Laws of Sines, Cosines, and Sides

In introductory physics, most of the trigonometry that you will use involves only right angles. However, in some situations, it may be easier to use one of the following rules that relates the side angles with angles for any triangle. Let a, b, and c be the sides of any triangle, and let A, B, and C be the three angles opposite these sides, respectively, as shown in Figure 5.8.

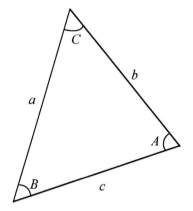

Figure 5.8

Then, according to the *law of sines*,

$$\frac{a}{\sin A} = \frac{b}{\sin B} = \frac{c}{\sin C}$$

According to the *law of cosines*,

$$a^2 = b^2 + c^2 - 2bc \cos A,$$
$$b^2 = a^2 + c^2 - 2ac \cos B,$$
$$c^2 = a^2 + b^2 - 2ab \cos C.$$

It is not difficult to remember the law of cosines. It is the Pythagorean theorem plus the cross term involving two side lengths and the cosine of the included angle.

According to the *law of sides*,

$$a = b \cos C + c \cos B,$$
$$b = a \cos C + c \cos A,$$
$$c = a \cos B + b \cos A.$$

Example 5.7.1

Mr. Johnson walks 2.0 km due north, then makes a 120° right turn and walks for 2.5 km along a straight line. How far is Mr. Johnson from the origin of the trip? What is the direction of his original location from his present position?

Assume Mr. Johnson starts from the point P. In this problem, $PQ = 2.0$ km, $QR = 2.5$ km, and angle $NQR = 120°$. Note that the triangle PQR is not a right triangle. The sides are marked as a, b, and c and the corresponding opposite angles are marked as A, B, and C. Determine PR ($= a$) and the angle B.

$A = 180° - 120° = 60°$

Use the law of cosines to get PR:

$$PR^2 = a^2$$

$$= b^2 + c^2 - 2bc \cos A.$$

Substitute the values

$$= (2.0 \text{ km})^2 + (2.5 \text{ km})^2 - 2(2.0 \text{ km})(2.5 \text{ km}) \cos 60°$$

$$= 4 + 6.25 - 5 = 5.25.$$

Take the square root

$$a = \pm \sqrt{5.25} = \pm 2.3 \text{ km.}$$

Take the positive root only

$a = 2.3$ km (since side a must be positive).

To determine the angle B, use the law of sines:

$$\frac{a}{\sin A} = \frac{b}{\sin B}.$$

Substitute the values

$$\frac{2.3 \text{ km}}{\sin 60} = \frac{2 \text{ km}}{\sin B}$$

or

$$\sin B = \frac{2. \text{ km}}{2.3 \text{ km}} \sin 60°$$

$$= \frac{2. \text{ km}}{2.3 \text{ km}} (0.87) = 0.76.$$

Take the *inverse sin* to get *B*

$$B = \text{INV} \sin (0.76) = 49.5°.$$

Exercises

Use the laws of sines, cosines, or sides
to determine the following:

27 The angle *A*
28. The angle *C*
29. The side *c*

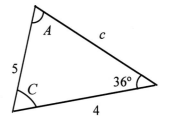

Answers to Exercises in Chapter 5

1. $h_a = 1.7$, $h = 3.5$

2. $h_o = 0.87$, $h = 1.73$

3. 20 cm

4. 7.7 cm

5. 713 m

22. $\cos^{-1}(0.56) = 55.9°$, $\tan^{-1}(2.5) = 68.2°$, $\sin^{-1}(0.92) = 66.9°$

23. 36.9°

24. 66.4°

25. $\sqrt{1-x^2}$, x, $\sqrt{1-x^2}/x$, $x/\sqrt{1-x^2}$, $1/\sqrt{1-x^2}$, $1/x$.

26. $x/\sqrt{1+x^2}$, $1/\sqrt{1+x^2}$, x, $1/x$, $\sqrt{1+x^2}/x$, $\sqrt{1+x^2}$

27. $A = 28°$

28. $C = 116°$

29. $c = 7.7$

Chapter 6
Vectors: Tracking the Direction of Quantities

Vectors are introduced in physics to represent certain physical quantities and to determine the direction of a change. The physical laws can be expressed in simple forms in vector notation. The different concepts of vector algebra are described.

Introduction to Vectors

6.1   Scalars and Vectors

Physical quantities are generally of two types:

- *Scalar*

- *Vector*

A quantity, such as the volume of a container, temperature of air, time interval between two events, distance of travel, and speed of a moving object, that can be expressed by a number and its unit is called a *scalar*. However, in some cases, a direction is also needed for the complete specification of a quantity. Examples include the displacement, velocity, and acceleration of an object. These are explained in physics texts and briefly described here.

Distance and Speed as Examples of Scalars

The distance of an object at any time is the total length of travel, and it is always positive. For a moving object, its distance changes with time. The time rate at which the distance of a moving object changes is called its *speed*. The *average speed* of a moving object is defined as

$$\text{average speed} = \frac{\text{distance traveled}}{\text{total time to travel the distance}} = \frac{d}{\Delta t} = \frac{d}{t_1 - t_2},$$

where d is the distance traveled and Δt ($= t_2 - t_1$) is the total time elapsed in traveling that distance. The SI unit for the speed is meters per second (m/s). Since *distance* is a *scalar* quantity, *speed* is also a *scalar*.

Displacement, Velocity, Acceleration, and Force as Examples of Vectors

If you walk 3 miles due north and 4 miles due east, you have traveled 7 miles but you are only 5 miles from your starting point [Figure 6.1 a]. That is, the magnitude of your displacement is 5 miles. The reason you cannot add in the ordinary way is that the displacement has *direction* and *magnitude*. The *displacement* of a moving object is the straight-line distance between its final position and its initial position, along with the *direction* from the starting position to the final position.

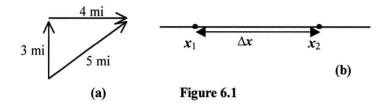

(a)   **Figure 6.1**   **(b)**

If the initial and final positions of an object are x_1 and x_2 at time t_1 and t_2, respectively, the displacement is [Figure 6.1b]

$$\Delta x = x_2 - x_1.$$

In one dimension, there are only two possible directions of motion (left or right). The x-axis is commonly used for horizontal motions and a plus (+) sign is used to indicate the direction to the right and a minus (−) sign is used to indicate the direction to the left. Displacement is expressed in meters and it can be positive or negative. Because of its direction, displacement is a *vector* quantity.

The *velocity* of a moving object tells how fast the object is moving *and* in what direction. The average velocity is the displacement divided by the total travel time:

$$\text{average velocity} = \frac{\text{displacement}}{\text{total travel time}} = \frac{\Delta x}{\Delta t} = \frac{x_2 - x_1}{t_2 - t_1}.$$

The SI unit of velocity is meters per second (m/s). Since velocity has a direction, it is a vector quantity.

The *velocity* of a moving object changes when the object speeds up (such as when you press the accelerator of your car) or slows down (such as when you step on the brake) or when the direction of motion of the car changes (such as when you make a turn). The time rate of change of velocity of an object is called its *acceleration*. The *average acceleration* is defined as the change in velocity divided by the time taken to make the change:

$$\text{average acceleration} = \frac{\text{change in velocity}}{\text{time to make the change}} = \frac{\Delta v}{\Delta t} = \frac{v_2 - v_1}{t_2 - t_1}$$

The SI unit of velocity is meters per second per second (m/s^2). Since velocity is a vector quantity, *acceleration* is also a *vector* quantity.

Force is simply a push or pull. It causes changes in motion or, in other words, causes an acceleration of an object.   Force (**F**) is related to acceleration (**a**) by Newton's second law of motion as: **F** = *m* **a**, where *m* is the mass of an object.   Since acceleration is a vector quantity, force is also a *vector* quantity.

Note: There are other vector quantities in physics.   Examples include linear momentum, angular momentum, torque, electric field, and magnetic field.

A *vector* is a physical quantity that has both a magnitude and a direction.   Since vectors and scalars are different types of physical quantities, they *cannot be* equal to each other. To distinguish them, different notations are used.   Scalars are represented by a single symbol in *italics*. **Boldface** is used to indicate a vector with a written symbol, and *italic* is used for its magnitude. An arrow is drawn to indicate a vector on a graph.   The length of the arrow is proportional to the magnitude of the vector, and the direction of the arrowhead specifies the direction of the vector.   Use arrows when writing vectors by hand.

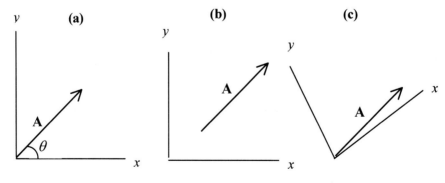

Figure 6.2

Figure 6.2 a shows a two-dimensional vector **A** whose magnitude is A and whose direction (as shown by the direction of the arrowhead) is specified by the angle θ made with the positive x-axis. The magnitude of a vector is always a positive number, but the vector can be pointed any direction, depending on the direction or the angle θ. Note that although a vector can be visualized on coordinate axes, it is independent of the coordinate system. In other words, the magnitude or direction of a vector does not change if you change the coordinate system or move the vector from one place to the other in the same coordinate system (Figure 6.2 b–c). Thus, you can move a vector anywhere keeping its magnitude and direction the same.

Since a vector is characterized by both its magnitude and direction, *two vectors* are said to be *equal* when both their magnitudes *and* directions are equal.

6.2   Components of a Vector

Rectangular Vector Components

By rectangular components of a vector we mean those at right angles to each other. Suppose that the two vectors **A** and **B** are at right angles. Then the magnitude of the resultant vector **C** (Figure 6.3 a) is given by the Pythagorean theorem:

$$C = \sqrt{A^2 + B^2} \, .$$

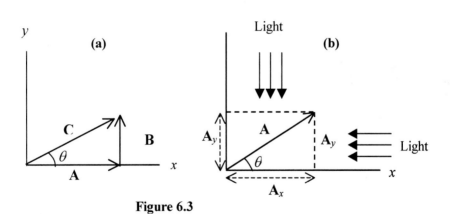

Figure 6.3

The orientation of **C** with respect to the x-axis is given by the inverse tangent relationship

$$\theta = \tan^{-1} \frac{B}{A} \, .$$

146   Chapter 6

In this case, **C** is the resultant vector and **A** and **B** are called the component vectors.

Resolving a Vector into Rectangular Components

Resolving a vector into rectangular components is the reverse of adding the components. To find the components of a vector, you need to set up a coordinate system, say, a Cartesian system. If a vector **A** makes an angle θ relative to the positive x-axis of the coordinate system [Figure 6.3b], the x and y components of the vector are the projections of the vectors on the x- and y-axes, respectively. If A_x and A_y are the x and y components, respectively, of **A,** then

$$A_x = A \cos \theta$$

and

$$A_y = A \sin \theta.$$

A_x and A_y can be positive, negative, or zero depending on θ. The sign of A_x and A_y in different coordinates are shown in Table 6.1.

Table 6.1

	Quadrant I	Quadrant II	Quadrant III	Quadrant IV
A_x	positive	negative	negative	positive
A_y	positive	positive	negative	negative

The Pythagorean theorem gives the precise value of the magnitude of the vector in terms of its components:

$$A = \sqrt{A_x^2 + A_y^2}$$

The direction of the vector can be obtained from the magnitude of the components and is given by

$$\theta = \tan^{-1} \frac{A_y}{A_x}.$$

Note: In determining an angle by means of the inverse tangent function, a calculator always displays the smallest angle. The displayed angle may or may not be the correct angle for a problem. Always utilize the quadrant grid to check your answer when you use the inverse tangent function. If you think the larger angle is the correct angle for the problem, add 180° to the smaller angle for the answer.

To **visualize** the vector components, imagine that a light beam is shining on the vector **A** from above the x-axis. Then the length of the shadow of the vector on the x-axis is the x component A_x of the vector. Similarly, if you imagine that a light beam is shining on the vector **A** from the right of the y-axis, the shadow of **A** on the y-axis is the y component A_y of the vector.

Example 6.2.1

Find the components of a vector of magnitude 16 m that is applied in a direction of 30° with the positive x-axis.

In this case, the vector **A** = 16 m and θ = 30°. The sketch of the problem follows. Use the relatioships $A_x = A \cos \theta$ and $A_y = A \sin \theta$ to determine the components A_x and A_y of **A**:

$$A_x = A \cos \theta = (16 \text{ m}) \cos 30°$$
$$= 16 \times 0.866 \text{ m} = 13.9 \text{ m}$$

$$A_y = A \sin \theta = \theta = (16 \text{ m}) \sin 30°$$
$$= 16 \times 0.5 \text{ m} = 8.0 \text{ m}$$

Example 6.2.2

A force (**F**) is a push or pull. It is a vector and is measured in the unit of newtons (N). The x and y components of a force are $F_x = -4$ N and $F_y = 6$ N. Find the magnitude and direction of the force.

We have
$$F_x = -4 \text{ N}$$
$$F_y = 6 \text{ N}.$$

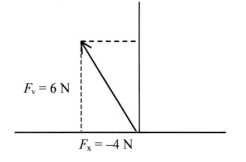

In this problem, since the x component is negative and the y component is positive, the force vector lies in the second quadrant. Hence, the angle θ is more than 90°.

Use the Pythogorean relation and the inverse tangent relation as mentioned in Section 6.2 to solve for the vector magnitude F and the angle θ, respectively:

$$F = \sqrt{F_x^2 + F_y^2}$$

Substitute the values of F_x and F_y

$$= \sqrt{(-4\text{N})^2 + (6\text{N})^2} = 7.2 \text{ N}$$

$$\theta = \tan^{-1}\frac{F_y}{F_x}$$

$$= \tan^{-1}\left(\frac{6\text{ N}}{-4\text{ N}}\right)$$

$$= -56.3° \text{ (using a calculator)}.$$

The calculator has given the smallest angle, which is in the fourth quadrant. Since the vector lies in the second quadrant, the correct angle is $180° + (-56.3°) = 123.7°$.

Exercises

1. The velocity of a moving object is the rate of change of its displacement with time, and it is a vector. A projectile is shot from the top of a tower with a velocity of 60 m/s at an angle of 32° with the horizontal direction. Determine the horizontal and vertical components of the velocity vector of the projectile.

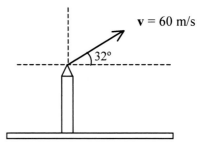

2. The x and y components of a force vector are $F_x = -2.0$ N and $F_y = -4.0$ N, respectively. Determine the force vector and its direction with the positive x-axis.

6.3   Unit Vectors

Unit vectors are introduced in mathematics and physics to express vectors in a convenient form. A unit vector is a dimensionless vector whose magnitude is unity and whose direction is usually one of the coordinate axes. There are *several kinds* of unit vectors in physics. Cartesian unit vectors and normal unit vectors are used in introductory physics and are described here.

Cartesian Unit Vectors

In most situations in introductory physics, you will use *Cartesian unit vectors*. Cartesian unit vectors are written as $\hat{\mathbf{i}}$, $\hat{\mathbf{j}}$, and $\hat{\mathbf{k}}$ or simply **i**, **j**, and **k**, pointing in the positive x-, y-, and z-axes, respectively. The unit vectors **i**, **j**, and **k** are mutually perpendicular and they meet at the origin as shown in Figure 6.4.

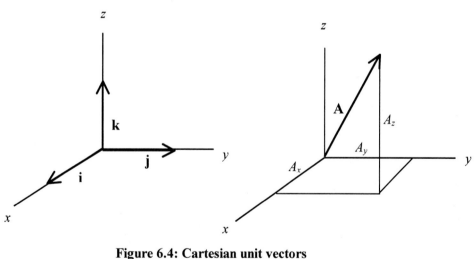

Figure 6.4: Cartesian unit vectors and Cartesian components of a vector

An arbitrary vector **A** in space can be written as the sum of its vector components using Cartesian unit vectors as follows:

$$\mathbf{A} = A_x\mathbf{i} + A_y\mathbf{j} + A_z\mathbf{k};$$ in two dimensions it reduces to

$$\mathbf{A} = A_x\mathbf{i} + A_y\mathbf{j}.$$

Normal Unit Vector

The normal unit vector is usually labeled as $\hat{\mathbf{n}}$ and it is pointed outward in a direction perpendicular to the surface at the point of interest [Figure 6.5]. If the surface is a plane, the normal unit vector is directed perpendicularly to the surface area. Use of the normal unit vector is common in electricity to calculate electric flux passing through an area. A small element of area Δs has some particular orientation with respect to the coordinate axes.

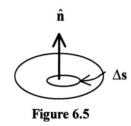

$\hat{\mathbf{n}}$

$\Delta \mathbf{s}$

Figure 6.5

A direction has been associated with such an area along the unit vector $\hat{\mathbf{n}}$ normal to the surface, and the area is considered to be a vector ($\Delta \mathbf{s} = \Delta s\,\hat{\mathbf{n}}$).

Vector Addition and Subtraction

6.4   Vector Addition

Vector quantities (such as displacement, velocity, and acceleration) can be added (*vector addition*) to obtain the *resultant vector*. Either the geometric method or the analytic component method can be used to add two or more vectors.

Vector Addition: Geometric Method

Either the triangle method or the parallelogram method can be used to add two vectors geometrically.

The Triangle Method or the Tip-to-Tail Method

To add two vectors **A** and **B** by the *triangle method*, first choose a suitable scale to represent the two vectors by two arrows on graph paper, so that the magnitudes of the vectors are proportional to the lengths of the arrows and the directions of the arrowheads show the directions of the vectors. Then, place the tail of vector **B** at the tip of vector **A**. The resultant vector (the vector sum) **R** is the vector joining the tail of **A** to the head of **B**, or **R** = **A** + **B** [Figure 6.6]. This method is also called the *tip-to-tail method*, since the tip of one vector is placed at the tail of the second vector.

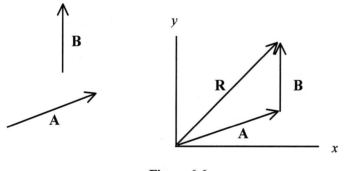

Figure 6.6

Measure the length of the resultant vector **R** (with a ruler) and the angle of the vector with the positive *x*-axis (using a protractor) for the direction. Change the length of the arrow to the magnitude of the vector by dividing by the chosen scale factor to get the magnitude of the resultant vector.

Parallelogram Method

Another graphical method, similar to the triangular method, is the *parallelogram method*. Draw the vectors **A** and **B** tail-to-tail, and form a parallelogram as shown in Figure 6.7. The diagonal represents the resultant vector **R** whose magnitude and direction can be measured from the diagram in a way similar to the triangular method.

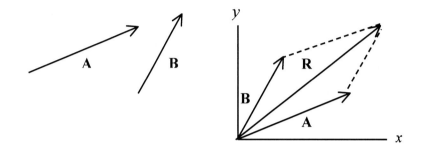

Figure 6.7

Polygon Method

The triangle method can be extended to add more than two vectors. Place all the vectors head-to-tail one by one, as shown in Figure 6.8, for the five vectors **A**, **B**, **C**, **D**, and **E**. The line joining the tail of the first vector **A** to the head of the last vector **E** represents the resultant vector. The method is called the polygon method, because the resulting figure is a polygon. In this case, **R** = **A** + **B** + **C** + **D** + **E**. The technique can be used to add any finite number of vectors.

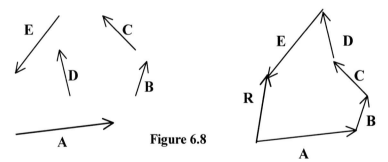

Figure 6.8

Vector Subtraction

An arrow having the same length as the original vector, but pointing in the opposite direction, represents the negative of a vector.

To subtract two vectors, **A** and **B**, add **A** to –**B**. The resultant vector **R** = **A** – **B** = **A** + (–**B**) (Figure 6.9).

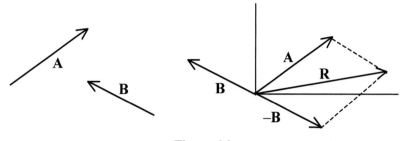

Figure 6.9

Even though the graphical methods of adding two or more vectors are useful and easy to understand, they are usually not very accurate.

Vector Addition Using Components

The *analytical component method* of vector addition involves resolving the vectors into components and then adding the components for each axis separately. If **A** and **B** are the two plane vectors to be added, and the resultant vector is **R** (= **A** + **B**), then [Figure 6.10]

$R_x = A_x + B_x$ and

$R_y = A_y + B_y.$

From the Pythagorean theorem,

$$R = \sqrt{R_x^2 + R_y^2},$$

and the angle is given by the inverse tangent relationship,

$$\theta = \tan^{-1}\left(\frac{R_y}{R_x}\right).$$

Figure 6.10

In the notation of a unit vector,

$$\mathbf{R} = \mathbf{A} + \mathbf{B} = (A_x\mathbf{i} + A_y\mathbf{j}) + (B_x\mathbf{i} + B_y\mathbf{j})$$

$$= (A_x + B_x)\mathbf{i} + (A_y + B_y)\mathbf{j}.$$

If both **A** and **B** have three components,

$$\mathbf{R} = \mathbf{A} + \mathbf{B} = (A_x\mathbf{i} + A_y\mathbf{j} + A_z\mathbf{k}) + (B_x\mathbf{i} + B_y\mathbf{j} + B_z\mathbf{k})$$

$$= (A_x + B_x)\,\mathbf{i} + (A_y + B_y)\,_y\mathbf{j} + (A_z + B_z)\,\mathbf{k}.$$

Similarly, $\mathbf{A} - \mathbf{B} = (A_x - B_x)\,\mathbf{i} + (A_y - B_y)\,\mathbf{j} + (A_z - B_z)\,\mathbf{k}.$

Vector addition is *commutative*: that is, $\mathbf{A} + \mathbf{B} = \mathbf{B} + \mathbf{A}$.
Vector addition is *associative*: that is, $(\mathbf{A} + \mathbf{B}) + \mathbf{C} = \mathbf{A} + (\mathbf{B} + \mathbf{C})$.

Example 6.4.1
Two vectors are given by $\mathbf{A} = 2\mathbf{i} + 3\mathbf{j} - 4\mathbf{k}$ and $\mathbf{B} = \mathbf{i} + 2\mathbf{j} + 3\mathbf{k}$. Determine the resultant vector $\mathbf{R} = \mathbf{A} + \mathbf{B}$.

We are given
$$\mathbf{A} = 2\mathbf{i} + 3\mathbf{j} - 4\mathbf{k}$$
$$\mathbf{B} = \mathbf{i} + 2\mathbf{j} + 3\mathbf{k}$$

$$\mathbf{R} = \mathbf{A} + \mathbf{B}$$

Substitute the values of **A** and **B** $= (2\mathbf{i} + 3\mathbf{j} - 4\mathbf{k}) + (\mathbf{i} + 2\mathbf{j} + 3\mathbf{k})$

Rearrange the terms $= (2 + 1)\mathbf{i} + (3 + 2)\mathbf{j} + (-4 + 3)\mathbf{k}$

$$= 3\mathbf{i} + 5\mathbf{j} - \mathbf{k}.$$

Example 6.4.2
A mail carrier leaves the post office and drives 20 mi due north to the next town. He then drives in a direction 30° north of east for 40 mi. Determine the displacement (**R**) of the mail carrier.

Let us choose the positive *x*-axis to be east and the positive *y*-axis north. Call the first displacement vector **A** and the second displacement vector **B**.

Since the first displacement **A** is in the *y* direction (north), it follows that

$A_x = 0.$

$A_y = 20$ mi.

Since the displacement **B** is 30° with the positive

x-axis (east), we have

$B_x = (40 \text{ mi}) \cos 30° = (40 \text{ mi}) (0.866) = 34.6$ mi.

$B_y = (40 \text{ mi}) \sin 30° = (40 \text{ mi}) (0.5) = 20$ mi.

The resultant displacement is **R**, as shown at

the right.

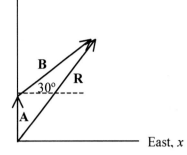

Add the x components to get R_x $R_x = A_x + B_x = 0 + 34.6 \text{ mi} = 34.6$ mi.

Add the y components to get R_y $R_y = A_y + B_y = 20 \text{ mi} + 20 \text{ mi} = 40$ mi.

Magnitude of R $R = \sqrt{R_x^2 + R_y^2} = \sqrt{(34.6 \text{ mi})^2 + (40 \text{ mi})^2} = 52.9$ mi

Angle θ $\theta = \tan^{-1}\left(\dfrac{R_y}{R_x}\right) = \tan^{-1}\left(\dfrac{40 \text{ mi}}{34.6 \text{ mi}}\right) = 49.1°.$

You can also solve the problem using unit vector notation.

In unit vector notation,   **A** = 0**i** + 20 mi **j**   **B** = 34.6 mi **i** + 20 mi **j**, and

     R = **A** + **B**

Subtitute the vectors **A** and **B**   = (0 + 34.6 mi) **i** + (20 mi + 20 mi) **j**

     = (34.6 mi) **i** + (40 mi) **j**.

Example 6.4.3
Relative Velocity: Raindrops are falling vertically with a velocity of 4 m/s. A bicyclist races through the rain with a horizontal velocity of 6 m/s relative to the ground. Determine the magnitude and direction of the velocity of the raindrops with respect to the bicyclist.

Since velocity is a vector quantity, the velocity of the raindrops with respect to the bicyclist is

$$\mathbf{v}_{RC} = \mathbf{v}_{RG} - \mathbf{v}_{CG},$$

where $\mathbf{v}_{RG}$ = velocity of the raindrop relative to the ground
and $\mathbf{v}_{CG}$ = velocity of the cyclist relative to the ground.

From the figure at the right, in unit vector notation,

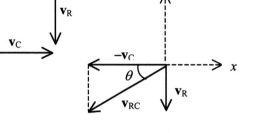

$$\mathbf{v}_{RC} = (-6 \text{ m/s}) \, \mathbf{i} + (-4 \text{ m/s}) \, \mathbf{j}.$$

The magnitude

$$v_{RC} = \sqrt{(-6 \text{ m/s})^2 + (-4 \text{ m/s})^2} = 7.2 \text{ m/s}.$$

The angle

$$\theta = \tan^{-1}\left(\frac{-4 \text{ m/s}}{-6 \text{ m/s}}\right) = 53.1°.$$

Exercises

3. Three vectors are given by $\mathbf{A} = \mathbf{i} - 3\mathbf{j} + 4\mathbf{k}$, $\mathbf{B} = 2\mathbf{i} + 4\mathbf{j} - \mathbf{k}$ and $\mathbf{C} = 3\mathbf{i} - \mathbf{j} + \mathbf{k}$. Determine $\mathbf{A} + \mathbf{B}$, $\mathbf{A} - \mathbf{B}$, $\mathbf{B} - \mathbf{A}$, $\mathbf{A} + \mathbf{B} + \mathbf{C}$, and $\mathbf{A} - \mathbf{B} + \mathbf{C}$.

4. A jogger runs 100 m in a direction 30° east of north (call it displacement vector $\mathbf{A}$) and then 50 m in a direction 60° south of east (call it displacement vector $\mathbf{B}$). Determine the resultant displacement $\mathbf{R}$ of the jogger. Express $\mathbf{A}$, $\mathbf{B}$, and $\mathbf{R}$ in unit vector notation.

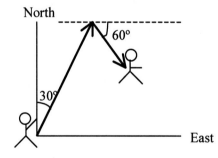

Vector Products

6.5   Products of Vectors

Multiplying a Vector by a Scalar

Multiplication of a vector by a positive scalar, s, multiplies the magnitude of the vector, but the direction of the vector does not change. If s is *negative*, the direction is *reversed*. Scalar multiplication is distributive. That is,

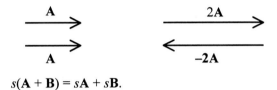

$s(\mathbf{A} + \mathbf{B}) = s\mathbf{A} + s\mathbf{B}.$

Note: When multiplying vectors in component form by a scalar, multiply each

component of the vector by the scalar.

Scalar Product or Dot Product

The *scalar product* of two vectors is defined as the *scalar* equal to the product of the magnitudes of the vectors and the cosine of the angle between them. That is, [Figure 6.11]

$\mathbf{A} \cdot \mathbf{B} = A\,B \cos\theta.$

Because of the notation, the
scalar product is also called the *dot product*.

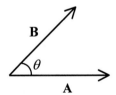

Figure 6.11

It is clear from the preceding equation that $\mathbf{A} \cdot \mathbf{B}$ can be positive, negative, or zero (zero when $\mathbf{A}$ and $\mathbf{B}$ are perpendicular, since $\cos 90° = 0$) depending on θ. Also, $\mathbf{A}^2 = \mathbf{A} \cdot \mathbf{A} = A\,A \cos 0° = A^2$. In unit vector notation,

$\mathbf{A} = A_x\mathbf{i} + A_y\mathbf{j} + A_z\mathbf{k}, \quad \mathbf{B} = B_x\mathbf{i} + B_y\mathbf{j} + B_z\mathbf{k},$ and

$$\mathbf{A} \cdot \mathbf{B} = (A_x\mathbf{i} + A_y\mathbf{j} + A_z\mathbf{k}) \cdot (B_x\mathbf{i} + B_y\mathbf{j} + B_z\mathbf{k}).$$

$\mathbf{i} \cdot \mathbf{j} = \mathbf{j} \cdot \mathbf{k} = \mathbf{k} \cdot \mathbf{i} = 0$ (since the angle between each pairs is 90°),

$\mathbf{i} \cdot \mathbf{i} = \mathbf{j} \cdot \mathbf{j} = \mathbf{k} \cdot \mathbf{k} = 1$ (since the magnitudes of $\mathbf{i}$, $\mathbf{j}$, and $\mathbf{k}$ are unity).

By direct multiplication of $\mathbf{A}$ and $\mathbf{B}$, $\mathbf{A} \cdot \mathbf{B} = A_xB_x + A_yB_y + A_zB_z$.

Note: To determine the dot product, multiply like components and then ad

The dot product is *commutative*, that is, $\mathbf{A} \cdot \mathbf{B} = \mathbf{B} \cdot \mathbf{A}$.
The dot product is *distributive*, that is, $\mathbf{A} \cdot (\mathbf{B} + \mathbf{C}) = \mathbf{A} \cdot \mathbf{B} + \mathbf{A} \cdot \mathbf{C}$.

Examples of Dot Products in Physics

Work as a Dot Product

A force can cause an object to move from one place to another. Consider the displacement of a particle of mass, m, under the influence of a force as shown in Figure 6.12. In this case, a component of the force acts along the direction of the displacement, and we say that *work* is done.
Quantitatively, we define work W by the force $\mathbf{F}$ in displacing an object by $\mathbf{r}$, as the component of the force causing the displacement   multiplied   by   the displacement. That is,

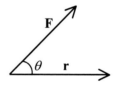

$$W = Fr \cos \theta = \mathbf{F} \cdot \mathbf{r}.$$

Figure 6.12

There are other examples of dot products in introductory physics. In electrostatics, you will find that the potential energy U of an electric dipole (two equal and opposite electric charges separated by a small distance) in an electric field $\mathbf{E}$ is given by $U = -\mathbf{p} \cdot \mathbf{E}$ (where $\mathbf{p}$ is called the electric dipole moment and $\mathbf{E}$ is the electric field), and, for a magnetic dipole, the potential energy in a magnetic field $\mathbf{B}$ is $U = -\mathbf{\mu} \cdot \mathbf{B}$. ($\mathbf{\mu}$ is the magnetic dipole moment.) In electricity and magnetism, the magnetic flux ϕ passing through an area A is given by the dot product $\phi = \mathbf{B} \cdot A\hat{\mathbf{n}}$.

Example 6.5.1

Determine the scalar product between vector $\mathbf{A} = 2\mathbf{i} + 4\mathbf{j} + 3\mathbf{k}$ and $\mathbf{B} = 4\mathbf{i} - 2\mathbf{j} + \mathbf{k}$. Find the angle between the two vectors.

$\mathbf{A} = 2\mathbf{i} + 4\mathbf{j} + 3\mathbf{k}$ The components of $\mathbf{A}$ are $A_x = 2$, $A_y = 4$, and $A_z = 3$.

$\mathbf{B} = 4\mathbf{i} - 2\mathbf{j} + \mathbf{k}$ The components of $\mathbf{B}$ are $B_x = 4$, $B_y = -2$, and $B_z = 1$.

To determine the angle θ between the vectors, use the definition $\mathbf{A} \cdot \mathbf{B} = A\,B\cos\theta$. Knowing $\mathbf{A} \cdot \mathbf{B}$, A, and B, one can determine the angle θ.

$$\mathbf{A} \cdot \mathbf{B} = A_xB_x + A_yB_y + A_zB_z.$$

Substitute the values of A_x, A_y, and A_z

$$= (2)(4) + (4)(-2) + (3)(1) = 3.$$

$$A = \sqrt{A_x^2 + A_y^2 + A_z^2}$$

Substitute the values of A_x, A_y, and A_z

$$= \sqrt{(2)^2 + (4)^2 + (3)^2} = 5.4.$$

$$B = \sqrt{B_x^2 + B_y^2 + B_z^2}$$

Substitute the values of B_x, B_y, and B_z

$$= \sqrt{(4)^2 + (-2)^2 + (1)^2} = 4.6.$$

$$\mathbf{A} \cdot \mathbf{B} = A\,B\cos\theta.$$

Subtitute the values of $\mathbf{A} \cdot \mathbf{B}$, A, and B $3 = (5.4)(4.6)\cos\theta$

or $\cos\theta = 0.12$

$$\theta = \cos^{-1}(0.12) = 83°.$$

Exercises

5. $\mathbf{A} = 2\mathbf{i} + 3\mathbf{j} - \mathbf{k}$, $\mathbf{B} = 4\mathbf{i} - 2\mathbf{j} + \mathbf{k}$, and $\mathbf{C} = \mathbf{i} + 3\mathbf{j} - 4\mathbf{k}$. Determine
    a) the dot product $\mathbf{A} \cdot \mathbf{B}$, b) the angle between $\mathbf{A}$ and $\mathbf{B}$, and c) $\mathbf{A} \cdot (\mathbf{B} + \mathbf{C})$.
    Also, show that $\mathbf{A} \cdot (\mathbf{B} + \mathbf{C}) = \mathbf{A} \cdot \mathbf{B} + \mathbf{A} \cdot \mathbf{C}$.

6. Work W is defined as the dot product of the force $\mathbf{F}$ and the displacement $\mathbf{r}$. That is, $W = Fr\cos\theta$. Find the work done in pulling a shopping cart by a force of 20 N at an angle of 30° by a distance of 2 m.

Vector Product or Cross Product

The vector product of two vectors **A** and **B** is written as **A X B** [Figure 6.13] and is also called a cross product. It is a vector perpendicular to both **A** and **B**, and its magnitude is defined as

$$|\mathbf{A} \times \mathbf{B}| = A\, B \sin\theta.$$

Geometrically, $|\mathbf{A} \times \mathbf{B}|$ is the area of the parallelogram generated by **A** and **B**.

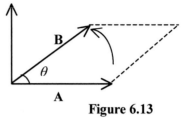

Figure 6.13

The direction of **A X B** (as shown in Figure 6.14) is given by the right-hand rule, which states that if the fingers of the right hand are curled in a way that will rotate **A** through the smaller angle into coincidence with **B**, then the thumb will point in the direction of **A X B**.

Thumb points in the direction of **A X B**

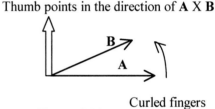

Curled fingers

Figure 6.14

From the definition, **A X A** = 0 and **A X B** = 0, if **A** and **B** are parallel.

Since the unit vectors **i**, **j**, and **k** are mutually perpendicular, and the cross product is perpendicular to both vectors, it follows that

> **i X i** = **j X j** = **k X k** = 0 (since sin 0° = 0).

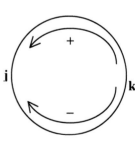

> **i X j** = **k**, **j X k** = **i**, **k X i** = **j**
>
> **j X i** = – **k**, **k X j** = – **i**, **i X k** = – **j**
>
> Follow the direction of the arrow in Figure 6.15 to determine the sign of the products.

Figure 6.15

$$\mathbf{A} \times \mathbf{B} = (A_x\mathbf{i} + A_y\mathbf{j} + A_z\mathbf{k}) \times (B_x\mathbf{i} + B_y\mathbf{j} + B_z\mathbf{k})$$

$$= (A_yB_z - A_zB_y)\ \mathbf{i} + (A_zB_x - A_xB_z)\ \mathbf{j} + (A_xB_y - A_yB_x)\ \mathbf{k}$$

This can be written as an easily remembered determinant:

$$\mathbf{A \times B} = \begin{vmatrix} \mathbf{i} & \mathbf{j} & \mathbf{k} \\ A_x & A_y & A_z \\ B_x & B_y & B_z \end{vmatrix}$$

Note: To determine the cross product, form the determinant whose first row is **i, j, k**, whose second row is the components of **A**, and whose third row is the components of **B**.

To get the first term, draw a horizontal line through the first row and a vertical line through the first column. Cross multiply the remaining four terms, obtaining $A_yB_z - A_zB_y$, and then multiply by **i**.

To get the second term, draw a horizontal line through the first row and a vertical line through the second column. Cross multiply the remaining four terms, obtaining $A_xB_z - A_zB_x$, and then multiply by $-\mathbf{j}$.

To get the third term, draw a horizontal line through the first row and a vertical line through the third column. Cross multiply the remaining four terms, obtaining $A_xB_y - A_yB_x$, and then multiply by **k**.

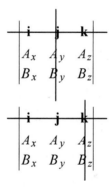

The order of multiplication is important: $\mathbf{A \times B} = -\mathbf{B \times A}$ (anticommutative law of vector multiplication).

The cross product is distributive: $\mathbf{A \times (B + C)} = \mathbf{(A \times B)} + \mathbf{(A \times C)}$.

Examples of Cross Products in Physics

There are several examples in physics where a physical quantity is defined as the cross product of two vectors.

Just as a force is needed to cause changes in linear motion, a *torque* is needed to cause changes in rotational motion. Torque is an example of a physical quantity that can be expressed as a cross product. *Torque* is defined as a vector given by $\tau = \mathbf{r} \times \mathbf{F}$ where $\mathbf{r}$ is the position vector of a point on an object and $\mathbf{F}$ is the force acting on the object at the point tending to rotate the object (Figure 6.16).

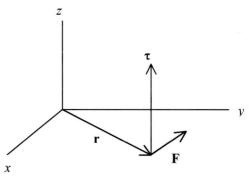

Figure 6.16

Since the position vector is defined with respect to a suitable point chosen as the origin, torque is defined about a point.

Angular momentum is another example of the cross product of two vectors. Just as an object in linear motion has a linear momentum $\mathbf{p} = m\mathbf{v}$ (where m is the mass and $\mathbf{v}$ is the velocity of the object), an object in rotational motion has a rotational momentum, called angular momentum. Angular momentum is defined as $\mathbf{L} = \mathbf{r} \times \mathbf{p}$.

Particles such as electrons and protons have electric charges. When an electric charge q moves with a velocity $\mathbf{v}$ in a magnetic field $\mathbf{B}$, it is found experimentally that there is a force on the moving charge. This is called magnetic force and its magnitude is found to be $F = qvB \sin \theta$.

The direction of the force, as observed experimentally, is perpendicular to the plane defined by $\mathbf{v}$ and $\mathbf{B}$ (Figure 6.17). Thus, the magnitude and direction of the force on a charged particle can be expressed as the cross product

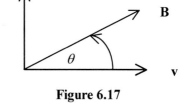

$$\mathbf{F} = q\,\mathbf{v} \times \mathbf{B}.$$

Figure 6.17

There are other examples of cross products in electricity and magnetism in introductory physics. A magnetic field exerts a force on an electric current. If the magnetic field is uniform over a straight wire of length $\mathbf{l}$, the force is $\mathbf{F} = I\,\mathbf{l} \times \mathbf{B}$, where I is the electric current. Also, a magnetic field tries to rotate a current loop. The torque

τ on a current loop in a magnetic field **B** is given by $\tau = \mu \times B$, where μ is called the magnetic dipole moment of the current loop, which is related to the number of turns, the current, and the area of the loop.

Example 6.5.1
Suppose the position vector is in the xy plane and is given by **r** = 1.8 m **i** + 1.1 **j**. If the force vector is **F** = 110 N **j**, determine the torque vector.

We have

$$\mathbf{r} = 1.8 \text{ m } \mathbf{i} + 1.1 \text{ } \mathbf{j} \qquad x = 1.8 \text{ m, } y = 1.1 \text{ m, } z = 0 \text{ m}$$

$$\mathbf{F} = 110 \text{ N } \mathbf{j} \qquad F_x = 0 \text{ N, } F_y = 110 \text{ N, } F_z = 0 \text{ N}$$

In determinant form,

$$\tau = \mathbf{r} \times \mathbf{F} = \begin{vmatrix} \mathbf{i} & \mathbf{j} & \mathbf{k} \\ A_x & A_y & A_z \\ B_x & B_y & B_z \end{vmatrix}.$$

$$= \begin{vmatrix} \mathbf{i} & \mathbf{j} & \mathbf{k} \\ x & y & z \\ F_x & F_y & F_z \end{vmatrix}$$

Substitute the values

$$= \begin{vmatrix} \mathbf{i} & \mathbf{j} & \mathbf{k} \\ 1.8 \text{ m} & 1.1 \text{ m} & 0 \\ 0 & 110 \text{ N} & 0 \end{vmatrix}.$$

Expand the determinant

$$= 0\mathbf{i} + 0\mathbf{j} + (1.8 \text{ m x } 110 \text{ N}) \mathbf{k}$$

$$= (180 \text{ m} \cdot \text{N}) \mathbf{k}.$$

Exercises

7. **A** = 2**i** + 3**j** – **k** and **B** = **i** –2**j** + 3**k**, determine **A** $\times$ **B**. Find the angle between **A** and **B**.

8. A particle is located at **r** = (2.0 m) **i** + (4.0 m) **j** +(3.0 m) **k**. A force **F** = (8 N) **j** – (1 N) **k** acts on the particle. Determine the torque about the origin.

9. A particle has a momentum $\mathbf{p}$ $(4\mathbf{i} + 6\mathbf{j})$ kg · m/s.   Find its angular momentum relative to the origin when it is at $\mathbf{r} = (2.0\,\mathbf{j} + 5.0\,\mathbf{k})$ m.

10. Show that $\mathbf{A} \cdot (\mathbf{B} \times \mathbf{C}) = \mathbf{C} \cdot (\mathbf{A} \times \mathbf{B}) = \begin{vmatrix} A_x & A_y & A_z \\ B_x & B_y & B_z \\ C_x & C_y & C_z \end{vmatrix}$.

Answers to Exercises in Chapter 6

1. x component = 50.9 m/s, y component = 31.8 m/s

2. F = 4.5 N, θ = 243°

3. $\mathbf{A} + \mathbf{B} = 3\mathbf{i} + \mathbf{j} + 3\mathbf{k}$, $\mathbf{A} - \mathbf{B} = -\mathbf{i} - 7\mathbf{j} + 5\mathbf{k}$, $\mathbf{B} - \mathbf{A} = \mathbf{i} + 7\mathbf{j} - 5\mathbf{k}$, $\mathbf{A} + \mathbf{B} + \mathbf{C} = 6\mathbf{i} + 4\mathbf{k}$, $\mathbf{A} - \mathbf{B} + \mathbf{C} = 2\mathbf{i} - 8\mathbf{j} + 6\mathbf{k}$

4. R = 86.6 m, θ = 30.0° north of east, $\mathbf{R}$ = 75.0 $\mathbf{i}$ + 43.3 $\mathbf{j}$

5 a) $\mathbf{A} \cdot \mathbf{B} = 1$ b) θ = 86.7°   c) $\mathbf{A} \cdot (\mathbf{B} + \mathbf{C}) = 16$

6. 34.6 N · m

7. $\mathbf{A} \times \mathbf{B} = 7\mathbf{i} - 7\mathbf{j} - 7\mathbf{k}$, angle θ = 59.9°

8. $-28\,\mathbf{i} + 2\mathbf{j} + 16\,\mathbf{k}$

9. $-30\mathbf{i} + 20\mathbf{j} - 8\mathbf{k}$

Chapter 7
How to Be Successful in a Physics Laboratory: Analysis of Data, Curve Fitting, Probability and Statistics

Statistical measurements are basically counts of the frequency of occurrence for different events in an experiment. Statistical analysis involves organizing the count data in the form of tables, graphing the data, and using the results to predict the future results. Data analysis is an important activity in statistics and in a physics laboratory. Let me start this chapter with some tips for success in a physics laboratory.

Experimental Error and Error Analysis

7.1 Introduction and Tips for Success in a Physics Laboratory

Physics laboratory is your chance to get hands-on experience and visually observe how physical principles are applied to real-world situations. The following tips should help you to enjoy your laboratory activities and improve your appreciation of laboratory physics:

- Be prepared before you come to the laboratory.

- Collaborate with other students in the group as well as other students in the lab.

- Think critically when you make any predictions or observations.

- Follow carefully the operational instructions and computer setup, if any.

- Take careful measurements.

- Analyze the data for the results. The analysis may involve making graphs or histograms or fitting curves (explained later in the chapter).

- Examine the results to draw a conclusion.

- Find experimental errors; that is, find the accuracy and the precision (explained later in the chapter) of your result wherever possible.

7.2   Experimental Data

In physics laboratory, you will set up experiments, take measurements, collect data, and analyze data to get results and conclusions. Experimental data are usually of two types:

- Analytical

- Statistical

When you measure a physical quantity as a function of one or more other known physical quantities, the collected data set is *analytical*. For example, if you record the position of a freely falling object as a function of time in an experiment, the collected data set is analytical. In an analytical data set, you usually use the data to discover any relationship between physical quantities (such as, how the position of a freely falling object depends on time).

But, when you record the frequency of occurrence of different events in an experiment, the collected data set is *statistical*. For example, suppose in an experiment you want to determine the energy distribution of the particles recorded in the experiment. In the experiment, you set up a small energy window and record how many particles have energy within this window. Then, you vary the window several times and, each time, determine the number again, thus generating the counts for the number of particles for different energy intervals. In this example, the collected data set is statistical.

7.3   Experimental Errors

Laboratory measurements of physical quantities always involve some experimental errors, regardless of your efforts to do experiments as carefully as possible. In other words, any real measurement will have some *error* or *uncertainty*. These errors could be any combination of the following:

- *Human error*

- *Instrumental error*

- *Computational error*

Human errors arise from personal bias or from carelessness in reading an instrument, in recording observations, or in mathematical calculations. *Instrumental error* can occur because of the bad calibration of an instrument or its limitations to record or measure accurately. *Computational errors* arise because no calculator or computer can calculate with infinite perfection and because of error propagation during different mathematical operations.

Experimental errors could be

- *Systematic*

- *Random*

Systematic errors are associated with particular instruments or techniques, such as an improperly calibrated instrument, incorrect use of the measuring device, or any bias on the part of the observer. One important source of systematic error is the *parallax*, which causes an apparent change in position due to the change in position of the eye—that is, looking at the measuring device from a wrong angle. For example, the length of the rod in Figure 7.1 may appear to be 8 or 10 units, respectively, if viewed above or below a line of sight perpendicular to the scale, although 9 units is the actual or true length. Systematic errors move the result in a single direction and cannot be minimized by repeating the measurements.

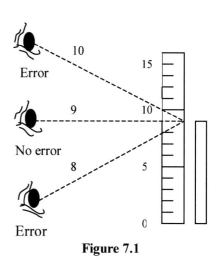

Figure 7.1

As another example of systematic error, think of an experiment that records the position of a freely falling object as a function of time and determines its velocity as a function of time and acceleration due to gravity *g* from the analysis of data. To determine *g*, it will be assumed that the velocity of the object increases linearly with time. The assumption holds for an ideal situation where friction and air drag are absent. In reality, the velocity would be less (although it could be negligibly small) because of these opposite effects. We can minimize systematic errors in an experiment by performing the experiment in a better environment (such as in a vacuum).

Random error results from unknown and unpredictable variations in experimental situations, such as temperature fluctuation of the environment. For random errors, the more measurements you make, the more errors will cancel each other. Thus, you can minimize random errors by repeating measurements many times and then taking the average of the measurements.

Since error cannot be totally avoided in a physics laboratory regardless of how good you are as an experimentalist, the analysis and reporting of errors is very important in a physics experiment.

7.4 Accuracy and Precision of Measurements: Mean or Average, Average Deviation, Variance, and Standard Deviation

You should understand the meaning of the terms *precision* and *accuracy* of a result that are commonly associated with the experimental measurements.

> **Note:** It is often said that *a measuring instrument that is high in precision is also high in accuracy.* This is not necessarily true! An instrument can provide precise results that could be way off from the actual result. For example, if you use an improperly calibrated ruler to measure a length carefully several times, your results can be precise, but not accurate.

Precision means the agreement among repeated measurements, that is, how close the individual results are. It is a measure of the magnitude of uncertainty of the result, without any reference to what the actual result is. For example, say two independent experiments of a time measurement provide the following results: 1.2 ± 0.1 seconds and 1.3 ± 0.2 seconds, respectively. The spread in the first measurement is between 1.1 and 1.3 seconds, whereas the spread in the second measurement is 1.1 and 1.5 seconds. Since the spread is less in the first measurement, the first measurement is more precise than the second measurement.

The *accuracy* of a measurement tells us how close the result is to the true or the accepted value. It is a measure of correctness of a result. The accepted value of a physical quantity can be found in the *CRC Physics Handbooks* or in physics textbooks. For example, suppose two independent measurements for the determination of π provide the results 3.141 and 3.147, respectively. Since the true value of π up to four significant figures is 3.142, the first result is more accurate since it is closer to the true value.

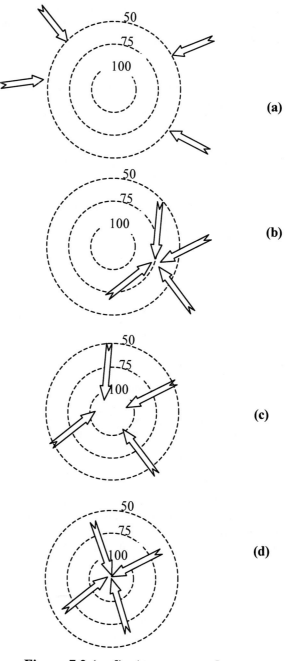

Figure 7.2 (a–d): Accuracy and Precision

To further clarify the basic difference between precision and accuracy, look at Figure 7.2 (a–d) for an analogy, where the true value is the bull's-eye. Figure 7.2 a has poor accuracy and poor precision. Figure 7.2b has poor accuracy, but good precision. Figure 7.2c has good accuracy, but poor precision. Figure 7.2d has good accuracy and good precision.

The *accuracy* of a measurement can be expressed by the *percentage error*:

$$\text{Percentage error} = \frac{|AcceptedValue - MeasuredValue|}{AcceptedValue} \times 100\,\% \,.$$

The *precision* of a measurement is expressed by calculating the *average deviation* or the *standard deviation* of the results.

The *average deviation* or *mean deviation* is the mean of absolute deviations of the individual results from the average value of the measurements. Thus, if x_1, x_2, ..., x_N are N individual measurements and $\bar{x}$ is their average value, then

$$\bar{x} = \frac{x_1 + x_2 + ... + x_N}{N} \,.$$

Absolute deviation of $x_1 = |x_1 - \bar{x}|$,

Absolute deviation of $x_2 = |x_2 - \bar{x}|$, etc.

Thus, the average deviation (a.d.) is given by

$$\text{a.d.} = \frac{|x_1 - \bar{x}| + |x_2 - \bar{x}| + ... + |x_N - \bar{x}|}{N}$$

$$= \frac{1}{N}\sum_{i=1}^{N}|x_i - \bar{x}| \,.$$

As the number of measurements is increased, the average of the set of measurements becomes more and more accurate in proportion to the square root of the number of measurements, N. This is reflected in the average deviation of the mean (A.D.) given by

$$\text{A.D} = \frac{\text{a.d.}}{\sqrt{N}} \,.$$

The A.D. represents the deviation of the arithmetical mean from the actual value and is known as the *probable error*. According to the theory of probability, A.D.

represents a 50% chance that the actual value of the quantity would lie between the A.D. of the mean.

The average of the square of the deviations of a set of measurements is called the *variance*, σ^2; that is,

$$\sigma^2 = \frac{\left(x_1 - \bar{x}\right)^2 + \left(x_2 - \bar{x}\right)^2 + ... + \left(x_N - \bar{x}\right)^2}{N}$$

$$= \frac{1}{N}\sum_{i=1}^{N}\left(x_i - \bar{x}\right)^2.$$

The *standard deviation* σ is defined as the square root of the variance, that is, the square root of the squares of the individual deviations. Thus,

$$\sigma = \sqrt{\frac{1}{N}\sum_{i=1}^{N}\left(x_i - \bar{x}\right)^2}$$

From the standpoint of the statistical theory, the standard deviation is the most important measure of dispersion, that is, the scatter of experimental points. When the number of measurements is small, it can be statistically shown that a better estimate of standard deviation is given by

$$\sigma = \sqrt{\frac{1}{(N-1)}\sum_{i=1}^{N}\left(x_i - \bar{x}\right)^2.}$$

The standard deviation is also called the *root mean square deviation* (or rms deviation), because you take the average of the squares of the deviations and then take the square root.

The *standard deviation of the mean* (SDOM) is given by

$$\text{SDOM} = \frac{\sigma}{\sqrt{N}}.$$

It is useful to know that when a large number of measurements of a quantity are made, the results form a *normal distribution* or *Gaussian distribution*, *p* (*x*), if the variations of the individual results occur due to random effects. A bell-shaped curve, symmetric about the average value of the measurements represents this distribution, which is shown in Figure 7.3. In this case, it can be statistically shown that there is a 68% chance (that is, 68% probability, in the language of statistics) that a measurement will fall within one standard deviation of the average value, $x = \pm \sigma$. The chance, or probability, that a measurement will fall within the two standard deviations of the average, $x = \pm 2\sigma$, can be shown to be 95%. See Appendix E for more details about Gaussian distribution.

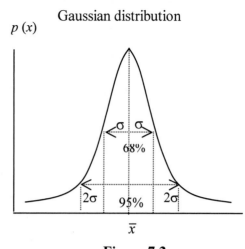

Gaussian distribution

p (x)

Figure 7.3

Example 7.4.1
In a series of experimental measurement for the value of the wavelength of a helium-neon laser light, the following results (in nanometers, 1 nanometer = 10^{-9} m) are obtained: 636.1, 641.2, 631.3, 645.1, 628.4, 624.8, 628.4, 627.8, 625.7, and 644.1.

Find the average value, $\bar{x}$, of the wavelength:
Find the percentage error of the measurement, where the accepted value of the wavelength is 632.8 nanometers. Find the average deviation (a.d.) of the results. Find the variance and standard deviation σ of the results. Also, find the A.D. and SDOM.

In this case, the average, $\bar{x} = \dfrac{x_1 + x_2 + ... + x_N}{N}$.

The number of measurements, $N = 10$. Therefore,

$$\bar{x} = \frac{636.1 + 641.2 + 631.3 + 645.1 + 628.4 + 624.8 + 628.4 + 627.8 + 625.7 + 644.1}{10} \text{ nm}$$

$= 633.3$ nm.

$$\text{Percentage error} = \frac{|AcceptedValue - MeasuredValue|}{AcceptedValue} \times 100\%$$

$$= \frac{|632.8 - 633.3|}{632.8} \times 100\% = 0.08\%$$

$$\text{Average deviation (a.d.)} = \frac{|x_1 - \bar{x}| + |x_2 - \bar{x}| + ... + |x_N - \bar{x}|}{N}$$

$$= \frac{1}{10}[|(636.1 - 633.3)| + |(641.2 - 633.3)| + |(631.3 - 633.3)| + |(645.1 - 633.3)| +$$

$$|(628.4 - 633.3)| + |(624.8 - 633.3)| + |(628.4 - 633.3)| + |(627.8 - 633.3)| +$$

$$|(625.7 - 633.3)|. + |(644.1 - 633.3)|]$$

$$= \frac{1}{10}[|2.8| + |7.9| + |-2.0| + |11.8| + |-4.9| + |-8.5| + |-4.9| + |-5.5| + |-7.6|. + |10.8|]$$

$$= \frac{1}{10}[2.8 + 7.9 + 2.0 + 11.8 + 4.9 + 8.5 + 4.9 + 5.5 + 7.6. + 10.8]$$

$$= 6.7 \text{ nm}$$

Thus, the wavelength = (633.3 ± 6.7) nm.
That is, the actual value of the wavelength has a high probability of lying between 640 nm (633.3 + 6.7 nm) and 626.8 nm (633.3 − 6.7 nm). Statistically, it can be shown that when the number of measurements is large, 57.5% of the individual results would lie within this interval (a.d.).

$$\text{Variance, } \sigma^2 = \frac{(x_1 - \bar{x})^2 + (x_2 - \bar{x})^2 + ... + (x_N - \bar{x})^2}{N}$$

$$= \frac{1}{10}\left[\begin{array}{l}(2.8)^2 + (7.9)^2 + (-2.0)^2 + (11.8)^2 + (-4.9)^2 + (-8.5)^2 + (-4.9)^2 + (-5.5)^2 + \\ (-7.6)^2 + (10.8)^2\end{array}\right]$$

$= 53.8 \text{ nm}^2$

Standard deviation $\sigma = \sqrt{53.8}$ nm $= 7.3$ nm

This means that there is a 68% chance that an individual measurement of the wavlength would lie within (633.3 ± 7.3) nm

$$\text{A.D} = \frac{\text{a.d.}}{\sqrt{N}} = \frac{6.7 \text{ cm.}}{\sqrt{10}} = 2.1 \text{ cm.}$$

This means that the chance that the true value of the wavelength lie within the interval (633.3 ± 2.1) nm is 50%.

$$\text{SDOM} = \frac{\sigma}{\sqrt{N}} = \frac{7.3 \text{ cm}}{\sqrt{10}} = 2.3 \text{ cm.}$$

7.5   Propagation of Error

Errors become larger, or, in other words, errors propagate when you combine different measurements and their errors. Therefore, you should pay attention when combining mathematically different measurements and their errors.

Let us call δx_1, δx_2 the errors (which may be the standard deviations of each measurement) for the measurements of the two variables, x_1, x_2 respectively. In other words, δx_1, δx_2 are the *uncertainties* of measurement x_1, x_2, respectively. Table 7.1 summarizes the results for the total error or total uncertainty in different mathematical operations of x_1, x_2. Use the appropriate formula from the table when you mathematically combine two different results with their individual errors. Note that the fractional error in x_1 is $\delta x_1/x_1$, and the percentage error in x_1 is $(\delta x_1/x_1)$ x 100%, etc.

Table 7.1: Expressions for Total Uncertainty in Mathematical Operations

Mathematical Operation	Results with Total Uncertainty
Addition	$(x_1 \pm \delta x_1) + (x_2 \pm \delta x_2) = (x_1 + x_2) \pm \sqrt{\delta x_1^2 + \delta x_2^2}$
Subtraction	$(x_1 \pm \delta x_1) - (x_2 \pm \delta x_2) = (x_1 - x_2) \pm \sqrt{\delta x_1^2 + \delta x_2^2}$
Multiplication	$(x_1 \pm \delta x_1) \cdot (x_2 \pm \delta x_2) \approx (x_1 x_2) \pm$ $(x_1 x_2)\sqrt{\left(\dfrac{\delta x_1}{x_1}\right)^2 + \left(\dfrac{\delta x_2}{x_2}\right)^2}$
Division	$\dfrac{(x_1 \pm \delta x_1)}{(x_2 \pm \delta x_2)} \approx \dfrac{x_1}{x_2} \pm \dfrac{x_1}{x_2}\sqrt{\left(\dfrac{\delta x_1}{x_1}\right)^2 + \left(\dfrac{\delta x_2}{x_2}\right)^2}$
Multiplication by a constant	$A(x + \delta x) = Ax \pm A\delta x$
Raised to a power	$(x \pm \delta x)^n = x^n \pm n x^{n-1} \delta x$

Example 7.5.1

The lengths of two rods were measured and found to be 60.2 ± 0.1 cm and 64.1 ± 0.2 cm, respectively.
a. When the rods are joined, find the length of the combined rod.
b. A bug is found to crawl along the first rod starting from its one end with a speed of 2.1 ± 0.3 cm/s. How long does it take to reach the other end of the rod?
c. Find the area of a circle with the first rod as its radius.
d. Find the area of the rectangle whose two sides have the same lengths as the rods.

a) To determine the length of the combined rod with uncertainty use the formula

$$(x_1 \pm \delta x_1) + (x_2 \pm \delta x_2) = (x_1 + x_2) \pm \sqrt{\delta x_1^2 + \delta x_2^2}.$$

Therefore, the length of the combined rod $= (60.2 + 64.1) \pm \sqrt{\left(0.1\right)^2 + \left(0.2\right)^2}$.

Simplify the result $= (124.3 \pm 0.2)$ cm

b) To determine the time, we divide the length x by the speed v.

Therefore,   time $= \dfrac{(x \pm \delta x)}{(v \pm \delta v)}$ $= \dfrac{x}{v} \pm \dfrac{x}{v} \sqrt{\left(\dfrac{\delta x}{x}\right)^2 + \left(\dfrac{\delta v}{v}\right)^2}.$

Substitute the values $= \dfrac{60.2}{2.1} \pm \dfrac{60.2}{2.1} \sqrt{\left(\dfrac{0.1}{60.2}\right)^2 + \left(\dfrac{0.3}{2.1}\right)^2}.$

Simplify $= (28.67 \pm 4.1)$ seconds.

Round off the result $= (29 \pm 4)$ seconds.

c) The area of the circle $A = \pi r^2 = \pi (60.2)^2 = 11400$ cm^2.

The uncertainty is given by $\dfrac{\delta A}{A} = 2\dfrac{\delta r}{r}$ [since, if $y = ax^n$, then $\dfrac{\delta y}{y} = n\dfrac{\delta x}{x}$].

$$\delta A = 2\frac{\delta r \cdot A}{r} = 2\frac{0.1 \times 11379.49}{60.2} = 37.8 \text{ cm}^2.$$

Thus, area $= (11379.49 \pm 37.8)$ cm^2.

Round off the result $= (11379 \pm 38)$ cm^2.

d) The area of the rectangle =

$$(x_1 \pm \delta x_1) \cdot (x_2 \pm \delta x_2) = (x_1 x_2) \pm (x_1 x_2)\sqrt{\left(\frac{\delta x_1}{x_1}\right)^2 + \left(\frac{\delta x_2}{x_2}\right)^2}.$$

Substitute the values $= (60.2)(64.1) \pm (60.2)(64.1)\sqrt{\left(\dfrac{0.1}{60.2}\right)^2 + \left(\dfrac{0.2}{64.1}\right)^2}.$

Simplify $= (3858.82 \pm 13.64)$ cm^2.

Round off the result $= (3859 \pm 14)$ cm^2.

Plots and Curve Fitting

7.6   Analysis and Plot of Data

In an experiment in physics, measurements are made of the variable(s) of interest, and the raw result of an experiment is a set of measurements, called the *data set*. Usually in an experiment, there are two kinds of data. The quantity that is varied in the experiment is called the *independent variable*, and the quantity that is measured is called the *dependent variable*. The collected data sets are then analyzed correctly to get the final result. The analysis usually includes organizing the data set in the form of a table or spreadsheet, making a graph, fitting the data to an appropriate equation or function, and determining the final result and errors in the measurements.

A *table* is a listing of variables with each variable in separate columns. In most situations, the independent variable is arranged, in increasing order of its magnitudes, in the first column. When you make a table, it should have an appropriate title, and the column heads must have the names of the variables with their proper units.

Note: A common mistake made by students is to write only the names of the variables in the column heads of a table without their units. Make sure to enter the proper units in parentheses as shown here, where "s" stands for seconds and "m" for meters:

Time (s)	Position (m)
1	4.9
2	19.5, etc.

In a *Cartesian graph*, you will plot the variables, usually with the independent variable along the *x*-axis and the dependent variable along the *y*-axis. Make sure to give a *title* to the graph and *label the variables with their units*. The scale of the axes should be such that the graph fills the entire sheet of a paper or a reasonable portion of it, as appropriate. For preliminary inspection, a scattered graph is usually alright, but for the final graph, connect the points by a smooth line. If any experimental error for any data point needs to be indicated on the graph, indicate the errors using proportionately sized vertical lines drawn at the points. These proportionately sized vertical lines drawn at different points on a graph are known as the *error bars*. Figure 7.4 a shows a scattered plot of *y* versus *x* with error bars. There may be situations where you need to extend the graph beyond the range of the collected data set; that is, you need to *extrapolate* the graph to find the result. It is common to extrapolate a graph so that it meets the *y*-axis to get the *y* intercept at $x = 0$, the value of which may be physically meaningful. Figure 7.4 b shows that the straight line has been extrapolated (shown as a broken line) to meet the *y*-axis. Similarly, in some situations where data sets have missing data, you need to guess the trend; that is, you need to *interpolate* the graph for the result.   Figure 7.4 c shows that the data have been interpolated (broken line) to complete the line, indicating how *y* changes with *x* for the whole range of *x*.

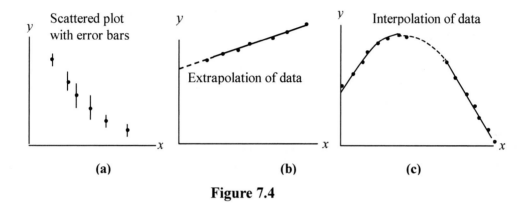

(a)   **(b)**   **(c)**

Figure 7.4

7.7   Semilog and Log–Log Graphs

A picture is worth a thousand words, and a well-constructed graph is worth at least five thousand, especially when it displays a large amount of information in a small space to make the patterns visible. Several coordinate systems and graph paper have been devised for graphical analysis and plotting.   In introductory physics, you will use mostly *rectangular*, or *Cartesian*, graphs, which are described in Section 4.4. On some occasions, you will need to plot logarithms of data.   In a rectangular Cartesian grid, the tick marks are equally spaced for both the *x*- and the *y*-axes.   If the *grid is logarithmic*, the tick marks are not equally spaced.   The spaces for numbers beginning with 1 and 2 are larger, compared with the numbers beginning with 8 or 9. There are two types of logarithmic graphs:

- *Semilog graph*

- *Log–log graph*

In semilog graphs, only the y-axis is logarithmic (i.e.,is, separation between ticks is proportional to the logarithm of numbers), while for the x-axis the ticks are equally spaced, as with for a rectangular Cartesian graph. For log–log graphs both the x- and the y-axes are logarithmic. Semilog and log–log grids are shown in Figures 7.5 a and b, respectively.

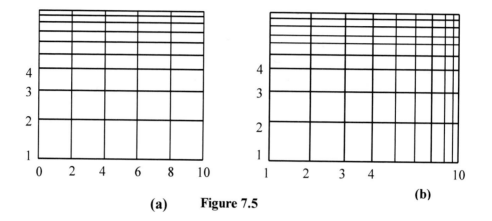

(a)   **Figure 7.5**   (b)

Semilog and log–log graphs have advantages over Cartesian graphs in some situations. For example, if the relationship between two physical quantities is $y = A\, e^{ax}$, a plot of x and y on a regular Cartesian graph gives an exponential curve, which is nonlinear. Taking the logarithm of the preceding equation, we get

$$ln\, y = ax + \ln A.$$

Thus, a plot of $ln\, y$ (or $log\, y$, since $ln\, y \approx 2.3\, log\, y$) versus x on a Cartesian graph, or, in other words, a plot of x and y on a semilog paper, gives a straight line with slope a and intercept $ln\, A$. For this reason, when the relationship between two physical quantities is exponential, a semilog graph is often used.

A plot $y = 2\, e^{3x}$ is shown in the Cartesian graph in Figure 7.6. Figure 7.7 shows the Cartesian plot of $ln\, y$ versus x for the equation $y = 2\, e^{3x}$. Note that the plot of $ln\, y$ versus x is a straight line with intercept ln 4 (= 0.60) and slope 3. (Vverify this by considering any two points on the graph!)

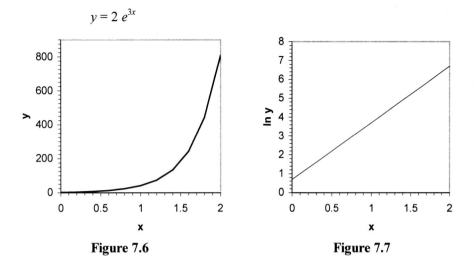

$y = 2\,e^{3x}$

Figure 7.6   **Figure 7.7**

If the relationship between two physical quantities x and y is a power function, $y = ax^n$ (where n is an integer other than 0 and 1), the graph is clearly a nonlinear curve in the Cartesian plot. If we take the logarithm of the power equation, $y = ax^n$, we get

$\ln y = \ln a + n \ln x$.

Thus, a plot of *log y* versus *log x*, or, in other words, a *log–log* plot of x and y, will produce a straight line with slope n and y-intercept *log a*. From the knowledge of the slope of the straight line, it is easy to determine the value of a and n. For this reason, a log–log graph is often used when the relationship is a power function.

A plot of $y = 4x^3$ is shown in the Cartesian graph in Figure 7.8. Figure 7.9 shows the corresponding log–log plot. Note that the plot is a straight line with intercept *ln* 4 and slope 3. (Verify this by considering any two points on the graph!)

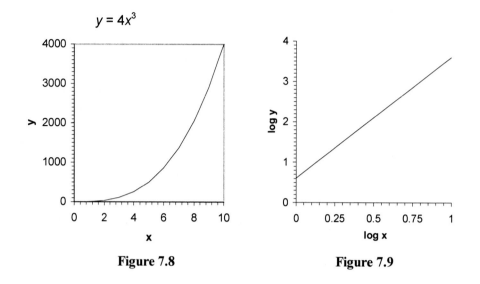

Figure 7.8

Figure 7.9

Exercise

7. Use the accompanying data to make a Cartesian plot and show that it is a nonlinear curve passing through the origin. Then, make a *log–log* plot, and show that it is a straight line. From the plots, discover that the relationship between the diameter D and volume V of a sphere is $V = 0.52\,D^3$. The data are as follows:

Diameter (m)	Volume (m^3)
0.10	0.00052
0.15	0.00177
0.20	0.00419
0.25	0.00818

7.8 Curve Fitting

In many situations, it is necessary to examine the plot to determine the mathematical function that describes the relationship between the variables plotted on the graph. The procedure of matching a graph of experimental variables with a known mathematical function is known as *curve fitting*. In curve fitting, you determine the best function that describes the relationship between the variables. Note that it is likely that the function may not pass through all the points of the graph.

In most situations in introductory physics, the fitted function is a straight line, a polynomial function of the independent variable, an exponential function, or a power function:

- A *linear* relationship between two variables x and y can be expressed by the equation $y = mx + b$, where m is called the slope of the line, defined in Chapter 4, and b is the y-intercept of the line (the y value where $x = 0$). Thus, the relationship is known when m and b are known. The purpose of the curve fitting is to determine the constant m and b if the relationship is a straight line.

- A *polynomial* function relationship between x and y is expressed by an equation $y = a_0 + a_1 x + a_2 x^2 + a_3 x^3 + \dots$, where $a_0, a_1, a_2, a_3, ..$ are constants to be determined from the curve fitting.

- An *exponential* relationship is expressed by $y = ae^{bx}$, where a and b are constants, to be determined from the curve fitting. Note that if you take the logarithm of the equation, it gives $\ln y = bx + \ln a$, which is a straight line if $\ln y$ is plotted along the y-axis with x along the x-axis.

- For a *power function*, the relationship is $y = ax^n$, where n is a positive integer and a is a constant to be determined from the curve fitting. If you take the logarithm on both sides, $\ln y = n \ln x + \ln a$. Thus, a plot of $\ln y$ versus $\ln x$ is a straight line.

> **Note:** If the relationship is exponential, make a semilog plot, and if the relationship is a power function, make a log–log plot, to find the coefficients.

The theory of curve fitting is not simple. The kind of mathematics and the level of difficulty depend on the shape of the fitted curve. *Least square fitting* fits data to a curve or a line so that the *square of the deviations of the data points from the fitted curve or line is the least*. If the data seem to fit to a straight line, least-square fitting, also called *linear regression*, is commonly used for the best-fitted line. It can be shown that the values of *m* and *b* for least-square fitting to a straight line can be determined from the expressions

$$m = \frac{N\left(\sum xy\right) - \left(\sum x\right)\left(\sum y\right)}{N\sum x^2 - \left(\sum x\right)^2}$$

and

$$b = \frac{\sum x^2\left(\sum y\right) - \left(\sum x\right)\left(\sum xy\right)}{N\sum x^2 - \left(\sum x\right)^2},$$

where *N* is the number of measurements (i.e., *N* is the number of pairs of *x* and *y* values).

You will generally be able to do the regression analyses of the experimental data using a standard calculator, computer spreadsheet, and other mathematics or physics software that you may use in your physics laboratory.

Example 7.8.1

A ball is dropped from a certain height. The accompanying data set represents the speed of the falling ball at different times:

Time (s)	Speed (m/s)
0.0	0.0
1.0	9.6
2.0	19.8
3.0	30.1
4.0	38.9
5.0	49.1
6.0	59.1
7.0	68.8

Make a plot of the best-fitted straight line from the data and determine the slope *m* and the *y*-intercept of the line.

The data are plotted to make a Cartesian graph of the speed of the object as a function of time. The best-fitted straight line which shows that the speed of the object increases linearly with time is shown in the following diagram:

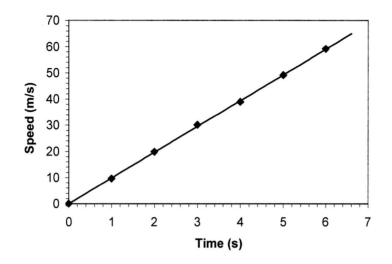

To determine the best-fitted line via linear regression, use the following expressions for the slope m and y intercept b:

$$m = \frac{N\left(\sum xy\right) - \left(\sum x\right)\left(\sum y\right)}{N\sum x^2 - \left(\sum x\right)^2}$$

$$b = \frac{\sum x^2\left(\sum y\right) - \left(\sum x\right)\left(\sum xy\right)}{N\sum x^2 - \left(\sum x\right)^2}.$$

Here N (the number of measurements) = 8. Thus,

$\sum x = (0 + 1 + 2 + 3 + 4 + 5 + 6 + 7)$ s $= 28$ s

$\sum x^2 = (0^2 + 1^2 + 2^2 + 3^2 + 4^2 + 5^2 + 6^2 + 7^2)$ s$^2 = 140$ s^2

$\sum y = (0 + 9.6 + 19.8 + 30.1 + 38.9 + 49.1 + 59.1 + 68.8)$ m/s $= 275.4$ m/s

$\sum xy = (0 \cdot 0 + 1 \cdot 9.6 + 2 \cdot 19.8 + 3 \cdot 30.1 + 4 \cdot 3.9 + 5 \cdot 49.1 + 6 \cdot 59.1 + 7 \cdot 68.8)$ m

$= 1376.8$ m

$$m = \frac{N\left(\sum xy\right) - \left(\sum x\right)\left(\sum y\right)}{N\sum x^2 - \left(\sum x\right)^2} = \frac{8\,(1376.8\text{ m}) - (28\text{ s})(275.4\text{ m/s})}{8 \times 140\text{ s}^2 - (28\text{ s})^2} = \frac{3303.2}{336}\text{ m/s}^2$$

$= 9.8$ m/s^2.

$$b = \frac{\sum x^2\left(\sum y\right) - \left(\sum x\right)\left(\sum xy\right)}{N\sum x^2 - \left(\sum x\right)^2} = \frac{140\,\text{s}^2\,(275.4\,\text{m/s}) - (28\,\text{s})(1376.8\,\text{m})}{8 \times 140\,\text{s}^2 - \left(28\,\text{s}\right)^2} = \frac{5.6}{336}\,\text{m/s}$$

= 0.02 m/s = 0.0 m/s (after rounding off to tenth's place)

The relationship is $y = (9.8\ \text{m/s}^2)\,x$.

Example 7.8.2

A mass attached to a light string suspended from a fixed support constitutes a simple pendulum. The time to complete one cycle of oscillation for the mass is called its period.   In a physics laboratory, the period of a simple pendulum is recorded for different lengths of the pendulum. The recorded data are as follows:

Length (m)	Period (s)
0.125	0.74
0.225	0.98
0.325	1.14
0.425	1.36
0.525	1.44
0.625	1.60
0.725	1.75
0.825	1.82
0.925	1.95
1.025	2.06

Make a scatter plot of the length of the pendulum along the x-axis and its period along the y-axis, and show that the relationship between the length and period for a simple pendulum is nonlinear.   Then, plot the length with the square of its period and show that the relationship is linear.   Do the regression analysis for the linear graph, and determine the slope and y-intercept of the line.

Following is the scatter plot of the data:

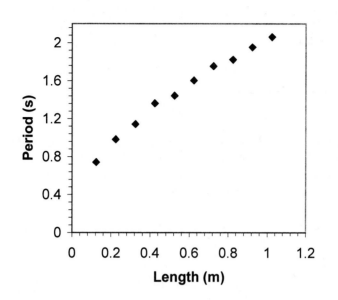

The relationship seems to be nonlinear.

From the raw data, square values of the periods are calculated, and the raw data, together with the square of the periods, are listed as follows:

Length (m)	Period (s)	Square of the period (s^2)
0.125	0.74	0.55
0.225	0.98	0.96
0.325	1.14	1.30
0.425	1.36	1.85
0.525	1.44	2.07
0.625	1.60	2.56
0.725	1.75	3.06
0.825	1.82	3.31
0.925	1.95	3.80
1.025	2.06	4.24

Following is the best-fitted solid straight line for the plot of length versus square of the period:

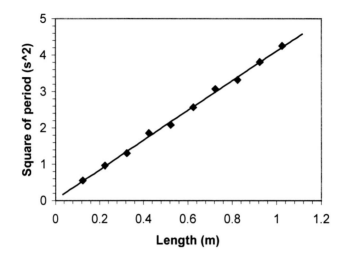

The slope and *y*-intercept of the line are calculated from the regression analysis.

Here, N (the number of measurements) = 10.

Σx = 0.125 + 0.225 + 0.325 + 0.425 + 0.525 + 0.625 + 0.725 + 0.825 + 0.925 + 1.025 = 5.75 m

Σx^2 = $(0.125)^2$ +$(0.225)^2$ + $(0.325)^2$ +$(0.425)^2$ + $(0.525)^2$ + $(0.625)^2$ + $(0.725)^2$ + $(0.825)^2$ + $(0.925)^2$ + $(1.025)^2$ = 4.13 m^2

Σy = 0.55 + 0.96 + 1.30 + 1.85 + 2.07 + 2.56 + 3.06 + 3.31 + 3.80 + 4.24 = 23.7 s^2

Σxy = 0.125 · 0.55 + 0.225 · 0.96 + 0.325 · 1.30 + 0.425 · 1.85 + 0.525 · 2.07 + 0.625 · 2.56 + 0.725 · 3.06 + 0.825 · 3.31 + 0.925 · 3.80 + 1.025 · 4.24

= 16.99 $s^2 \cdot$ m

$$m = \frac{N\left(\Sigma xy\right) - \left(\Sigma x\right)\left(\Sigma y\right)}{N\Sigma x^2 - \left(\Sigma x\right)^2} = \frac{10\left(16.99\,s^2\,m\right) - \left(5.75\,m\right)\left(23.7\,s^2\right)}{10 \times 4.13\,m^2 - \left(5.75\,m\right)^2} = \frac{33.62}{8.23}\ s^2/m$$

= 4.08 s^2/m

$$b = \frac{\Sigma x^2\left(\Sigma y\right) - \left(\Sigma x\right)\left(\Sigma xy\right)}{N\Sigma x^2 - \left(\Sigma x\right)^2}$$

$$= \frac{4.13\,\mathrm{m}^2\left(23.7\,\mathrm{s}^2\right) - (5.75\,\mathrm{m})\left(16.99\,\mathrm{s}^2\,\mathrm{m}\right)}{10 \times 4.13\,\mathrm{m}^2 - \left(5.75\,\mathrm{m}\right)^2} = \frac{0.19}{8.24}\,\mathrm{s}^2 = 0.02\ \mathrm{s}^2.$$

Therefore, the relationship is $T^2 = (4.08\ \mathrm{s}^2/\mathrm{m})\ L + 0.02$, where L and T represent the length and period of the pendulum, respectively. Since 0.02 is negligible, the equation reduces to

$$T^2 = (4.08\ \mathrm{s}^2/\mathrm{m})\ L$$

or $$T = \left(2.02\ \mathrm{s}/\sqrt{\mathrm{m}}\right)\sqrt{L}.$$

Probability and Statistics

7.9   Probability

The concept of probability is important in physics and mathematics, and especially in statistical physics. Introductory wave mechanics, which represents the behavior of subatomic particles such as electrons and protons, treats the wave equation as an expression for the mathematical probability.

The probability of an occurrence tells us the chance or likelihood of an occurrence. For example, when we toss a coin, there are two possible outcomes: "Heads" (H) and "Tails" (T). If we toss the coin a large number of times, it is found that the chance, or the probability, of obtaining H or T is ½.

If an event occurs in a ways and fails to occur in b ways, and each of these events is equally likely, then the probability of occurring the event is $a/(a+b)$.
Thus, the mathematical probability is defined as the *ratio of a certain number of events to the total number of events.*

The mathematical probability is a fraction that is always less than or equal to unity. When the probability attains its maximum value of unity, it becomes a certainty. By definition, the total probability for obtaining all states is unity. Probability is expressed as a fraction, decimal, or percentage. A probability of ½ can be written as ½, 0.5, or 50%.

Example 7.9.1

Two coins are simultaneously tossed on a table. Make a list of the different possible outcomes and the probability of each outcome.

The possible outcomes and their numbers are listed in the following table:

Number of Heads	Possible Outcomes	Number of Outcomes
2	HH	1
1	HT, TH	2
0	TT	1

Total number of possible outcomes = 1 + 2 + 1 = 4.

Thus,

the probability of 2 heads $= \dfrac{1}{4}$;

the probability of 1 head $= \dfrac{2}{4} = \dfrac{1}{2}$; and

the probability of 0 head (2 tails) $= \dfrac{1}{4}$.

Note that total probability for all outcomes $= \dfrac{1}{4} + \dfrac{1}{2} + \dfrac{1}{4} = 1$, as expected.

Example 7.9.2

Four coins are simultaneously tossed on a table. Make a list of the different possible outcomes and the probability of each outcome.

The possible outcomes and their numbers are listed in the following table:

Number of Heads	Possible Outcomes	Number of Outcomes
4	HHHH	1
3	HHHT, HHTH, HTHH, THHH	4
2	HHTT, HTHT, HTTH, THTH, TTHH, THHT	6
1	HTTT, THTT, TTHT, TTTH	4
0	TTTT	1

Total number of possible outcomes $= 1 + 4 + 6 + 4 + 1 = 16$.

Thus,

the probability of 4 heads (that is, 0 tail) $= \dfrac{1}{16}$;

the probability of 3 heads (that is, 1 tail) $= \dfrac{4}{16} = \dfrac{1}{4}$;

the probability of 2 heads (that is, 2 tails) $= \dfrac{6}{16} = \dfrac{3}{8}$;

the probability of 1 head (3 tails) $= \dfrac{4}{16} = \dfrac{1}{4}$; and the probability of 0 head (that is, 4 tails) = $\dfrac{1}{16}$.

The total probability = $\dfrac{1}{16} + \dfrac{1}{4} + \dfrac{3}{8} + \dfrac{1}{4} + \dfrac{1}{16}$ = 1, as expected.

Probability of a composite event

Multiply individual probabilities to obtain sequential probabilities. Note that when only one coin is tossed, the probability of obtaining a head is ½. As found in Example 6.8.1, when two coins are simultaneously tossed, the probability of obtaining two heads is ¼, which is ½ X ½, the product of individual probabilities of obtaining each head separately. Similarly, when four coins are simultaneously tossed, the probability of obtaining four heads is 1/16 which is ½ X ½ X ½ X ½.

> Thus, if two events are mutually independent and P_1 is the probability of event 1 and P_2 is the probability of event 2, then the probability of both events occurring simultaneously is given by
> $$P_{12} = P_1 \text{ X } P_2.$$

7.10 Frequency Distribution, Histogram, and Probability Distribution

In a statistical measurement, the counts of certain events are recorded. Some events occur more frequently than others. In the language of statistics, this means that some events are more probable than others. You get a *frequency distribution* of different counts by plotting the number of counts for an event along the y-axis and the value of the events along the x-axis. When the frequency distribution is plotted in the form of a bar graph, it is called a *histogram*. In a histogram, the height of each bar in the graph is proportional to the frequency of events. The width of the bar is called the *bin*. Each bin in a histogram is the same so that we can compare the bars with each other.

It is common to plot a statistical data set in the form of a *probability distribution* instead of in the form of a frequency distribution or histogram. To get a probability distribution, the frequency of occurrence of an event or a count is first divided by the total number of events. If f is the frequency of a measurement and N is the total number of measurements, then the probability of occurrence of a measurement is $p = f/N$. In the graph, p is then plotted along the y-axis, the event along the x-axis. The resultant graph is usually shown as a histogram when the number of bins is not large. When the number of bins is large, usually the points are joined by a smooth line. This line represents the *limiting probability distribution* of the measurements of the events.

Example 7.10.1
In a get-together of boys and girls, the ages of 800 boys and girls were recorded in years and months. Make a histogram of the data.

The data consist of 800 different records. The *bin* size is chosen to be 1 year for this histogram; that is, the data were sorted so that ages between 10.5 years 11.4 years will fall into one category of 11 years, etc.   The frequency distribution and the corresponding histogram are as follows:

Age (years)	Frequency, f	Age (years)	Frequency, f
1	0	11	155
2	0	12	14
3	1	13	94
4	1	14	48
5	2	15	22
6	6	16	8
7	20	17	2
8	42	18	1
9	100	19	0
10	150	20	0

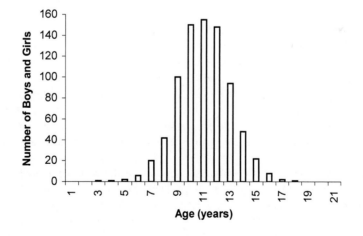

Example 7.10.2

For the previous example of a get-together of boys and girls, where the ages of 800 boys and girls were recorded in years and months, make a probability distribution of the data in form of a bar graph and a continuous line graph.

In general, to determine a probability distribution, you need to divide the frequency distribution by the total number of measurements or records. In this example, the total number N of boys and girls = 800. For example, the probability that a boy or girl will be age 10 is $p = f/N = 150/800 = 0.1875$. In a similar manner, determine the probability for each age. The resulting probability distribution data are as follows:

Age (years)	Probability, p	Age (years)	Probability, p
1	0	11	0.19375
2	0	12	0.185
3	0.00125	13	0.1175
4	0.00125	14	0.06
5	0.0025	15	0.0275
6	0.0075	16	0.01
7	0.025	17	0.0025
8	0.0525	18	0.00125
9	0.125	19	0
10	0.1875	20	0

The data are plotted in the form of a bar graph with the x-axis as age in years and the y-axis as probability:

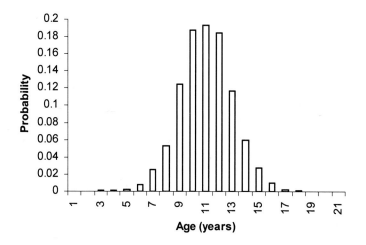

The data points are joined to generate a line graph that shows the probability distribution curve:

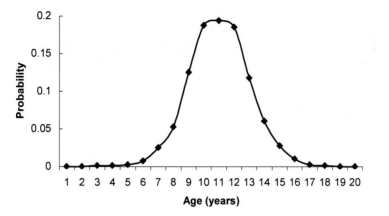

Note that, if the number of records were very large, the probability distribution line graph would be a smoother curve or function, which would be the limiting probability distribution.

7.11   Mean, Median, and Mode of a Distribution

In addition to determining the frequency distribution, it is often necessary or desirable to mathematically characterize the distribution in terms of different parameters that describe its center, width, shape, etc. The three commonly used parameters that measure the location of the center are *mean*, *median* and *mode*.

The mean, or average, of a distribution, as defined in Section 7.4, is the sum of all measurements divided by the number of measurements:

$$\bar{x} = \frac{\sum x}{N}.$$

The mean can also be determined by using the frequency distribution f or the probability distribution p, as follows:

$$\bar{x} = \frac{\sum fx}{N}$$

$$\bar{x} = \sum px.$$

Since each measurement is weighted by its frequency or probability, this technique of determining mean or average is also known as *weighted average*.

When the data of a measurement are sorted from lowest to highest, the *median* of the measurement represents the middle point of the data set. If there is an even number of data points, then it is the average of the two middle points. In situations where, because of many outlying measurements, the mean may be unreliable, the median is used.

The *mode* of a data set represents the measurement that occurs most often in the data set.

Note: *Mean* is the sum of the data divided by the number of data.

Median is the middle number when the data are arranged in numerical order.

Mode is the number(s) or item(s) that occurs most often. Remember the mnemonic: "*Mode occurs most often and median is in the middle.*"

Example 7.11.1

The ages (in years) of 15 students in a class are 17, 19, 17, 20, 22, 24, 28, 18, 19, 20, 24, 18, 18, 21, 23. Determine the mean, mode, and median for the ages.

The total number of students, N, is 15. First, use the formula $\bar{x} = \dfrac{\sum x}{N}$ to determine the mean:

$$\text{Mean} = \frac{1}{15}\left[17+19+17+20+22+24+28+18+19+20+24+18+18+21+23\right] \text{ years}$$
$$= 20.5 \text{ years.}$$

You can also use the formula $\bar{x} = \sum px$ to determine the mean. To do so, find the frequency of occurence (f) for each age and its corresponding probability ($p = f/N$). The results are as follows:

Age	17	18	19	20	21	22	23	24	25	26	27	28
f	2	3	2	2	1	1	1	2	0	0	0	1
p	0.13	0.20	0.13	0.13	0.07	0.07	0.07	0.13	0.00	0.0	0.0	0.07

Thus, the mean age is

$\bar{x} = \sum px = 17 \times 0.13 + 18 \times 0.20 + 19 \times 0.13 + 20 \times 0.13 + 21 \times 0.07 + 22 \times 0.07 +$

$23 \times 0.07 + 24 \times 0.13 + 25 \times 0.0 + 26 \times 0.0 + 27 \times 0.0 + 28 \times 0.07 = 20.5$ years.

To determine the *mode*, look at the data to find which age or ages appear most often; that is, find the age for which the frequency f (and the probability p) is a maximum. It is clear that f and p are maximum for the age 18. Therefore, the mode is 18 years.

To determine the *median* age, first list the 15 ages in order of increasing numbers. Then, find the middle number.

17   17   18   18   18   19   19   20   20   21   22   23   24   24   28

←——————————→ ⋀ ←——————————→
　　　7 numbers　　median　　7 numbers

The middle number is 20. Therefore, the median age is 20 years.

Measures of the Shapes of a Distribution

The general shape of a curve or a distribution is characterized by the following two quantities:

- Skewness

- Kurtosis

Figure 7.10　　　　　　　　　　　　**Figure 7.11**

The *skewness* of a distribution describes how much a curve is weighted to the left or to the right. This is explained in Figure 7.10. Skew right means that the tail of the curve is toward right, and skew left means the tail of the curve is toward left.

The *kurtosis* of a distribution is a measure of how sharp the peak of the distribution is. A distrbution with high kurtosis has a pointy peak, while a distribution with low kurtosis has a flat peak [Figure 7.11].

Appendix A

Important Conversion Factors

Length
1 mi =1.609 km
1 ft = 0.3048 m
1 m = 3.281 ft
1 in = 0.0254 m
1 nautical mile = 1.852 km
1 fermi = 10^{-15} m
1 angstrom = 10^{-10} m
1 light-year = 9.46×10^{15} m
1 parsec = 3.09×10^{16} m

Mass
1 kg = 0.0685 slug
1 atomic mass unit (u) = 1.6605×10^{-27} kg

Time
1 day = 8.64×10^4 s
1 year = 3.156×10^7 s

Angle
1 radian (rad) = 57.3°
1° = 0.01745 rad

Area
1 in^2 = 6.45×10^{-4} m^2
1 ft^2 = 9.29×10^{-2} m^2

Volume
1 liter (L) = 1000 mL = 1000 cm^3 = 1.0×10^{-3} m^3
1 gallon = 4 qt = 3.78 L

1 m^3 = 35.31 ft

Speed
1 mi/h = 1.609 km/h = 0.447 m/s
1 ft/s = 0.305 m/s
1 km/h = 0.278 m/s
1 knot = 1.151 mi/h = 0.5144 m/s
1 rev/m = 0.1047 rad/s

Force
1 lb = 4.45 N
1 N = 0.225 lb
1 N = 10^5 dyne

Work and Energy
1 J = 10^7 ergs = 0.738 ft·lb
1 ft·lb = 1.36 J
1 kcal = 4.18×10^3 J = 3.97 Btu
1 kWh = 3.60×10^6 J
1 electron volt (eV) = 1.6022×10^{-19} J

Power
1 W = 0.738 ft·lb/s
1 hp = 550 ft·lb/s = 746 W

Pressure
1 atm = 1.013×10^5 N/m^2
1 lb/in^2 = 6.90×10^3 N/m^2
1 Pa = 1 N/m^2 = 1.45×10^{-4} lb/in^2

Temperature
$$F = \frac{9}{5} C + 32$$
$$C = \frac{5}{9} (F - 32)$$
$$K = C + 273.15$$

Appendix B

Fundamental Physical Constants

It has been found that certain numbers play important roles in describing the universe and its physical principles. These numbers are called *fundamental constants*. Following is a list of most fundamental constants, their symbols, and their values.

Constants	Symbol	Value
Speed of light	c	2.9979×10^8 m/s
Gravitational constant	G	6.6726×10^{-11} N·m^2/kg^2
Planck's constant	h	6.6261×10^{-34} J·s
$h/2\pi$	$\hbar$	1.05457×10^{-34} J·s
Boltzmann's constant	k	1.3807×10^{-23} J/K
Molar gas constant	R	8.3145 J/mol·K
Avogadro's number	N_A	6.0214×10^{23} mol^{-1}
Stefan-Boltzman constant	σ	5.6705×10^{-8} /m^2·K^4
Permittivity of free space	ε_0	8.8542×10^{-12}/N·m^2
Permeability of free space	μ_0	1.2566×10^{-6} T·m/A
Charge of an electron	e	1.6022×10^{-19} C
Electron rest mass	m_e	9.1094×10^{-31} kg = .000549 u = 0.511 MeV/c^2
Proton rest mass	m_p	1.6726×10^{-27} kg = 1.00728 u = 938.3 MeV/c^2
Neutron rest mass	m_n	1.6750×10^{-27} kg = 1.008665 u = 939.6 MeV/c^2
Proton-electron mass ratio	m_p/m_e	1836.1527
Fine-structure constant, $\mu_0 c e^2/2h$	α	7.2974×10^{-3}
Inverse fine structure constant	α^{-1}	137.0360
Rydberg constant	R_∞	1.0974×10^{-7} m^{-1}
Faraday constant, $N_A e$	F	9.6485×10^{-4} C·mol^{-1}
Magnetic flux quantum, $h/2e$	Φ_0	2.06783×10^{-15} Wb

Appendix C

Useful Data and Values of Some Numbers

Useful Data

Absolute Zero (0 K) = –273.15 C

Earth: Mass = 5.97×10^{24} kg
Radius (mean) = 6.38×10^{6} m
Acceleration due to gravity (mean) on the surface = 9.8 m/s^2

Sun: Mass = 1.99×10^{30} kg
Radius (mean) = 6.96×10^{8} m

Moon: Mass = 7.35×10^{22} kg
Radius (mean) = 1.74×10^{6} m
Acceleration due to gravity = 1.67 m/s^2

Earth-sun distance (mean) = 149.6×10^{9} m
Earth-moon distance (mean) = 384×10^{6} m

Values of Some Numbers

$\pi = 3.1415927$
$e = 2.7182818$

$\ln 2 = 0.6931472$
$\log_{10} e = 0.4342945$
$\ln 10 = 2.3025851$

$\sqrt{2} = 1.4142136$
$\sqrt{3} = 1.7320508$

Appendix D

Solving System of Linear Equations: Method of Determinant (Cramer's Rule)

Cramer's rule provides a way for solving any set of n independent linear equations with n unknown variables. The procedure involves the calculation of the *determinants* of a *matrix*.

A matrix is a rectangular array of numbers or functions, such as,
$$\begin{pmatrix} a_{11} & a_{12} & a_{13} & a_{14} \\ a_{21} & a_{22} & a_{23} & a_{24} \\ a_{31} & a_{32} & a_{33} & a_{34} \end{pmatrix}.$$

The preceding matrix contains three rows (the number of horizontal lines in the array) and four columns (the number of vertical lines in the array). The *determinant* is defined only for square matrices, so that there are equal numbers of rows and columns. For example, for a matrix of order 2, $A = \begin{bmatrix} a_{11} & a_{12} \\ a_{21} & a_{22} \end{bmatrix}$, the determinant of the matrix is denoted as $|A|$ and is given by

$$|A| = \begin{vmatrix} a_{11} & a_{12} \\ a_{21} & a_{22} \end{vmatrix} = a_{11}\,a_{22} - a_{12}\,a_{21}.$$

In general, for a determinant of order n (that is, for a determinant containing n rows and n columns), the *minor* of an element a_{ij} of the determinant is defined as the determinant of order $(n-1)$ that is obtained by deleting the ith row and the jth column (these are the row and column containing a_{ij}) from the original determinant. The minor of a_{ij} is denoted as $|M_{ij}|$. The *cofactor* of a_{ij}, usually denoted as Δ_{ij}, is defined as

$$\Delta_{ij} = (-1)^{i+j}|M_{ij}|.$$

For example, for the determinant $\begin{vmatrix} a_{11} & a_{12} & a_{13} \\ a_{21} & a_{22} & a_{23} \\ a_{31} & a_{32} & a_{33} \end{vmatrix}$,

$$\Delta_{11} = (-1)^{1+1}|M_{11}| = + \begin{vmatrix} a_{22} & a_{23} \\ a_{32} & a_{33} \end{vmatrix}.$$

The value of a determinant $|A|$ of order n is given by the sum of the n products obtained by multiplying each element in one row or column by its cofactor. As an example, expand the following determinant:

$$|A| = \begin{vmatrix} 2 & 6 & 2 \\ 3 & 8 & 1 \\ 4 & 5 & 6 \end{vmatrix} = (-1)^{1+1} \, 2 \begin{vmatrix} 8 & 1 \\ 5 & 6 \end{vmatrix} + (-1)^{1+2} \, 6 \begin{vmatrix} 3 & 1 \\ 4 & 6 \end{vmatrix} + (-1)^{1+3} \, 2 \begin{vmatrix} 3 & 8 \\ 4 & 5 \end{vmatrix}$$

$$= 2(48 - 5) - 6(18 - 4) + 2(15 - 32)$$
$$= 86 - 84 - 34 = -32.$$

Note that, although there are several possible ways of expanding a determinant, it can be shown that all these yield the same result. (Verify this.)

Now, consider any set of linear equations, for example, three linear equations with three unknowns:

$$a_{11}x + a_{12}y + a_{13}z = c_1$$
$$a_{21}x + a_{22}y + a_{23}z = c_2$$
$$a_{31}x + a_{32}y + a_{33}z = c_3.$$

The equations can be written in matrix form as

$$\begin{pmatrix} a_{11} & a_{12} & a_{13} \\ a_{21} & a_{22} & a_{23} \\ a_{31} & a_{32} & a_{33} \end{pmatrix} \begin{pmatrix} x \\ y \\ z \end{pmatrix} = \begin{pmatrix} c_1 \\ c_2 \\ c_3 \end{pmatrix}$$

The solutions of x, y, and z, as given by Cramer's rule, are

$$x = \frac{\begin{vmatrix} c_1 & a_{12} & a_{13} \\ c_2 & a_{22} & a_{23} \\ c_3 & a_{32} & a_{33} \end{vmatrix}}{D} \qquad\qquad y = \frac{\begin{vmatrix} a_{11} & c_1 & a_{13} \\ a_{21} & c_2 & a_{23} \\ a_{31} & c_3 & a_{33} \end{vmatrix}}{D}$$

$$z = \frac{\begin{vmatrix} a_{11} & a_{12} & c_1 \\ a_{21} & a_{22} & c_2 \\ a_{31} & a_{32} & c_3 \end{vmatrix}}{D} \, , \qquad \text{where } D = \begin{vmatrix} a_{11} & a_{12} & a_{13} \\ a_{21} & a_{22} & a_{23} \\ a_{31} & a_{32} & a_{33} \end{vmatrix} \neq 0$$

The solution can be generalized for any number of equations and unknowns. If the denominator D is zero, the equations are not independent and Cramer's rule cannot be applied.

Cramer's rule is very useful in electricity to solve for electric currents in different sections of multiloop circuits.

Example (from electricity)
The electric currents (in milliampere, mA) i_1, i_2, and i_3 in different sections of a multiloop electric circuit are given by

$$i_1 + i_2 + i_3 = 0$$
$$2i_1 - i_2 + i_3 = -1$$
$$8i_1 - i_2 - 2i_3 = 11$$

Determine the currents.

Use Cramer's rule to solve for the currents. Comparing these equations with the preceding three linear equations, we find that the solutions are

$$i_1 = \frac{\begin{vmatrix} 0 & 1 & 1 \\ -1 & -1 & 1 \\ 11 & -1 & -2 \end{vmatrix}}{D} \, , \qquad i_2 = \frac{\begin{vmatrix} 1 & 0 & 1 \\ 2 & -1 & 1 \\ 8 & 11 & -2 \end{vmatrix}}{D} \, , \text{ and, } \quad i_3 = \frac{\begin{vmatrix} 1 & 1 & 0 \\ 2 & -1 & -1 \\ 8 & -1 & 11 \end{vmatrix}}{D} \, .$$

In this case,

$$D = \begin{vmatrix} 1 & 1 & 1 \\ 2 & -1 & 1 \\ 8 & -1 & -2 \end{vmatrix} = (-1)^{1+1} 1 \begin{vmatrix} -1 & 1 \\ -1 & -2 \end{vmatrix} + (-1)^{1+2} 1 \begin{vmatrix} 2 & 1 \\ 8 & -2 \end{vmatrix} + (-1)^{1+3} 1 \begin{vmatrix} 2 & -1 \\ 8 & -1 \end{vmatrix}$$

$$= 1(2+1) - (-4-8) + (-2+8) = 3 + 12 + 6 = 21.$$

$$\begin{vmatrix} 0 & 1 & 1 \\ -1 & -1 & 1 \\ 11 & -1 & -2 \end{vmatrix} = (-1)^{1+1} 0 \begin{vmatrix} -1 & 1 \\ -1 & -2 \end{vmatrix} + (-1)^{1+2} 1 \begin{vmatrix} -1 & 1 \\ 11 & -2 \end{vmatrix} + (-1)^{1+3} 1 \begin{vmatrix} -1 & -1 \\ 11 & -1 \end{vmatrix}$$

$$= 0 - (2 - 11) + (1 + 11) = 9 + 12 = 21.$$

$$\begin{vmatrix} 1 & 0 & 1 \\ 2 & -1 & 1 \\ 8 & 11 & -2 \end{vmatrix} = (-1)^{1+1} 1 \begin{vmatrix} -1 & 1 \\ 11 & -2 \end{vmatrix} + (-1)^{1+2} 0 \begin{vmatrix} -1 & 1 \\ 11 & -2 \end{vmatrix} + (-1)^{1+3} 1 \begin{vmatrix} 2 & -1 \\ 8 & 11 \end{vmatrix}$$

$$= (2 - 11) + 0 - (22 + 8) = -9 + 30 = 21.$$

$$\begin{vmatrix} 1 & 1 & 0 \\ 2 & -1 & -1 \\ 8 & -1 & 11 \end{vmatrix} = (-1)^{1+1} 1 \begin{vmatrix} -1 & -1 \\ -1 & 11 \end{vmatrix} + (-1)^{1+2} 1 \begin{vmatrix} 2 & -1 \\ 8 & 11 \end{vmatrix} + (-1)^{1+3} 0 \begin{vmatrix} 2 & -1 \\ 8 & -1 \end{vmatrix}$$

$$= (-11 - 1) - (22 + 8) + 0 = -12 - 30 = -42.$$

Therefore, the solutions are

$$i_1 = \frac{\begin{vmatrix} 0 & 1 & 1 \\ -1 & -1 & 1 \\ 11 & -1 & -2 \end{vmatrix}}{D} = \frac{21}{21} = 1 \text{ mA}, \quad i_2 = \frac{\begin{vmatrix} 1 & 0 & 1 \\ 2 & -1 & 1 \\ 8 & 11 & -2 \end{vmatrix}}{D} = \frac{21}{21} = 1 \text{ mA},$$

$$i_3 = \frac{\begin{vmatrix} 1 & 1 & 0 \\ 2 & -1 & -1 \\ 8 & -1 & 11 \end{vmatrix}}{D} = \frac{-42}{21} = -2 \text{ mA}.$$

Appendix E
Gaussian Distribution

Four important types of limiting distributions that are often found in physical problems or experiments are the *Gaussian distribution, Poisson distribution, exponential distribution*, and *binomial distribution*. The *Gaussian distribution* is the most common distribution among these distributions. It is characterized by a symmetric bell-shaped curve and is also called *normal distribution* or *error distribution*, as shown in Figure 7.3. If a large number of random, independent variables are measured in an experiment, the results form a Gaussian distribution symmetric with respect to the mean value of the result. The Gaussian distribution is characterized by the mathematical expression

$$p(x) = \frac{1}{\sqrt{2\pi\sigma^2}} e^{-(x-\bar{x})^2/2\sigma^2},$$

σ = Width parameter $\bar{x}$ = Center (or mean) of the distribution.

Example
When the error is random, the probability distribution of the result of a large number of measurements in an experiment is found to follow a Gaussian distribution $p(x) = \left(1/\sqrt{2\pi\sigma^2}\right) e^{-(x-\bar{x})^2/2\sigma^2}$, with average $\bar{x}$ = 12.0 units and width parameter σ = 2.2 units. Determine the probability that the result of a measurement x is i) 11.0 units and ii) 18.0 units.

In this case, $\bar{x}$ = 12.0 units, and the width parameter σ = 2.2 units.

Therefore, $p(x) = \dfrac{1}{\sqrt{2\pi\pi(2.2^2}} e^{-(x-12.0)^2/2(2.2)^2}$

When x = 11.0 units, $p(10) = \dfrac{1}{\sqrt{2\pi(2.2))^2}} e^{-(11.0-12.0)^2/2(2.2)^2} = 0.16.$

When x = 18.0 units $p(18) = \dfrac{1}{\sqrt{2\pi(2.2)^2}} e^{-(18.0-12.0)^2/2(2.2)^2} = 0.0044.$

Bibliography

R. David Gustafson and Peter D. Frisk, *Beginning and Intermediate Algebra 2nd: An Integrated Approach.* Brooks/Cole Publishing Company, Pacific Grove, CA, 1999, ISBN 0-534-35943-4.

Kristen A. Hubbard and Debora M. Katz, *The Physics ToolBox: A Survival Guide for Introductory Physics.* Brooks/Cole Publishing Company, Pacific Grove, CA, ISBN 0-03-034652-5.

Arnold D. Picker, *Preparing for General Physics.* Addison-Wesley Publishing Company, Reading, MA, 1992, ISBN 0-201-56952-3.

James B. Seaborn, *Mathematics for the Physical Sciences.* Springer, NY, 2001, ISBN 0-387-95342-6.

David R. Lide (ed.), *CRC Handbook of Chemistry and Physics.* 77th edition, CRC Press, NY, 1996, ISBN 0-8493-0477-6.

Robert C. Weast (ed.), *CRC Handbook of Tables for Mathematics.* 4th edition, CRC Press, NY.

M. E. Shanks, C. F. Brumfiel, C. R. Fleenor and R. E. Eicholz, *Pre-Calculus Mathematics.* Addison-Wesley Publishing Company, Reading, MA, 1968.

Douglas C. Giancoli, *Physics for Scientists and Engineers.* 3rd ed., Prentice Hall, Upper Saddle River, NJ, 2000, ISBN 0-13-021517-1.

Eugene Hecht, *Physics: Calculus.* 2nd ed., Brooks/Cole Publishing Company, Pacific Grove, CA, 2000, ISBN 0-534-36270-2.

William Faissler, *An Introduction to Modern Electronics.* John Wiley, NY, 1991, ISBN 0-471-62242-7.

Raymond A. Serway, *Physics for Scientists and Engineers.* 3rd ed., Saunders College Publishing, Philadelphia, 1992, ISBN 0-03-096027-4.

John D. Cutnell and Kenneth W. Johnson, *Physics.* 5th ed., John Wiley, NY, 2001, ISBN 0-471-32146-X.

James Walker, *Physics.* Prentice Hall, Upper Saddle River, NJ, 2002, ISBN 0-13-633124-6.